The Farm Labor Problem

The Farm Labor Problem

A Global Perspective

J. Edward Taylor

Diane Charlton

ACADEMIC PRESS

An imprint of Elsevier

Academic Press is an imprint of Elsevier
125 London Wall, London EC2Y 5AS, United Kingdom
525 B Street, Suite 1650, San Diego, CA 92101, United States
50 Hampshire Street, 5th Floor, Cambridge, MA 02139, United States
The Boulevard, Langford Lane, Kidlington, Oxford OX5 1GB, United Kingdom

Library of Congress Cataloging-in-Publication Data
A catalog record for this book is available from the Library of Congress

British Library Cataloguing-in-Publication Data
A catalogue record for this book is available from the British Library

ISBN 978-0-12-816409-9

For information on all Academic Press publications visit
our website at https://www.elsevier.com/books-and-journals

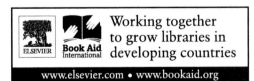

Working together
to grow libraries in
developing countries

www.elsevier.com • www.bookaid.org

Publisher: Susan Dennis
Acquisition Editor: Nancy Maragioglio
Editorial Project Manager: Ruby Smith
Production Project Manager: Nilesh Kumar Shah
Cover Designer: Mark Rogers

Typeset by SPi Global, India

Contents

For more information, visit - https://farmlabor.ucdavis.edu/.

Author Biography

J. Edward Taylor is a professor of Agricultural and Resource Economics at the University of California, Davis. He is a fellow of both the Agricultural and Applied Economics Association (AAEA) and the American Association for the Advancement of Science (AAAS). Ed coauthored the award-winning book *Beyond Experiments in Development Economics: Local Economy-wide Impact Evaluation* (Oxford University Press, 2014), *Essentials of Development Economics* (University of California Press, 2015), *Essentials of Applied Econometrics* (University of California Press, 2016), and *Worlds in Motion: Understanding International Migration at the End of the Millennium* (Oxford University Press, 2005). He is listed in *Who's Who in Economics* as one of the world's most cited economists. Ed has advised numerous foreign governments and international development agencies on matters related to economic development, and he has been an editor of the *American Journal of Agricultural Economics*.

Diane Charlton is an assistant professor in the Department of Agricultural Economics & Economics at Montana State University. Her research primarily focuses on the economics of rural labor markets and education, labor migration, agriculture, and development.

Preface

Every now and then you get a new course to teach and you have to write the book for it. Agricultural labor markets differ from other labor markets in fundamental ways, and they evolve differently than other labor markets as economies develop. When Ed took over the UC Davis Agricultural Labor course in 2015, with Diane as the lead teaching assistant, there were labor economics textbooks out there but nothing on agricultural labor. For the first offering of the course, we used handouts and an article or two and then wrote summary papers to fill in the gaps. As we continued our teaching, Ed at UC Davis and Diane at Montana State University, these diverse materials evolved into the chapters of this book. The end product weaves together economic analysis and the history of agricultural labor markets, using data and real-world events. The farm labor history of California and the United States is particularly rich, so it plays a central role in the book, but the book, like the course, has a global perspective. The challenges of keeping workers in the fields are as relevant to Europe and high-income Asian countries as to the United States. The changing role of farm labor during the agricultural transformation is as evident in China as in Mexico.

The chapters that follow provide the basics to understand how farm labor markets work (farm labor demand and supply, agricultural labor market equilibria over time and across space, and labor in agricultural household models); farm labor and immigration policy; farm labor organizing; farm employment and rural poverty; unionization and the United Farm Workers movement; the Fair Foods Program as a new approach to collective bargaining; the declining immigrant farm labor supply; and what economic development in relatively low-income countries portends for the future of agriculture in the United States and other high-income countries. The book concludes with a chapter called "Robots in the Fields," which extrapolates current trends to a perhaps not-so-distant future in which machines, endowed with advanced information technology, will harvest many or most crops.

Our goal in writing this book was to provide a basic reference to anyone interested in farm labor, including students, policy makers, agricultural producers and producer organizations, farm labor advocates, and rural development practitioners. Technical material appears in chapter appendices. Combining the text and appendices makes this an ideal textbook (indeed, the only textbook available) for undergraduate or graduate courses or course segments on agricultural labor.

We have many people to thank. Philip Martin has been a valuable mentor to both of us, and he offered critical feedback throughout this project. No one understands how farm labor markets and farm labor and immigration policy work better than Phil does. Antonio Yúnez-Naude has been a dear friend and collaborator, without whom our research in rural Mexico would not have happened. Richard Mines and Douglas Massey taught Ed the challenges and intricacies of doing rural economic fieldwork in Mexico and California (sometimes interviewing the same workers in both places). Daniel Carroll answered every question we had about the US National Agricultural Workers Survey (NAWS), and Dan Sumner proved to be the ultimate resource on agricultural producers and agricultural production economics. Steven Zahniser and Thomas Hertz have been invaluable collaborators, particularly in thinking through the implications of the end of farm labor abundance for policy makers. Aaron Smith, Steve Boucher, Katrina Jessoe, and Giovanni Peri provided expert advice on econometric strategies to identify the end of farm labor abundance and its causes, and Stavros Vougioukas provided a biological and engineering perspective on how farmers are adapting to it. A dream team of talented past and current UC Davis graduate students provided research assistance on farm labor projects and feedback as teaching assistants in the agricultural labor course. They include Dawn Thilmany (now a professor at Colorado State University), Mateusz Filipski (now an assistant professor at University of Georgia), Alejandro López-Feldman (now a professor at CIDE), Dale Manning (now an assistant professor at Colorado State University), and Anubhab Gupta, Heng Zhu, Zacharia Rutledge, Karen Ortiz, Huang Chen, Tomoe Bourdier, and Hairu Lang (all current UC Davis PhD students).

This book has many components drawn from a large number of research projects. Those projects received financial support from the Giannini Foundation of Agricultural Economics, the William and Flora Hewlett Foundation, the US Department of Agriculture (USDA), the National Institute of Food and Agriculture (NIFA), Mexico's Consejo Nacional de Ciéncia y Tecnología, and the United Nations Food and Agricultural Organization (UN-FAO).

Above all, Ed thanks his wife, Peri Fletcher, and sons, Sebastian and Julian Fletcher-Taylor, for their unwavering support and insights, both anthropological and scientific. Diane thanks her parents, Mark and Susan Charlton, for their continual prayers and support, and Ángeles Morales and Elihu García for their hospitality, friendship, and invaluable assistance while doing fieldwork in Mexico.

Chapter 1

Introduction

The fight is never about grapes or lettuce. It is always about people.

Cesar Chavez

Agriculture is different from other economic sectors of the economy in ways that have far-reaching implications for the analysis of labor markets (Timmer, 1988). A multitude of farms are scattered across a vast geographic space. Production and labor demands are seasonal and uncertain, separated in time and dependent on the whims of nature. Farmers make their planting decisions months (and in the case of perennial crops, years) ahead of harvest. Labor and other inputs have to be available exactly when farmers need them, sometimes with little to no advance notice. Perishable crops not harvested on time can rot in the fields. Late harvests can cause farmers to miss key marketing windows or contract deadlines, and crops left in the field are exposed to disease and weather risks. If labor is not available to plant crops and apply inputs on time, or if the rains do not come, there might not be anything to harvest at all. All of these considerations make the demand for farm labor uncertain and dynamic—that is, it changes over time.

The *supply* of farm workers is also uncertain and dynamic. As countries develop and their per-capita incomes rise, their workforces move out of agriculture and into nonfarm jobs, creating the specter of farm labor scarcity. Farmers in rich countries pressure their governments to open the doors to workers from poorer countries to fill the void. Even if there is an overall abundance of farm workers, local labor shortages materialize if workers are not available in the right place at the right time. Crops ripen unevenly across space, creating demand for follow-the-crop migrant workers. History shows that people do not choose to do follow-the-crop migration if they have other options.

This chapter introduces the complex challenges associated with farm labor demand and supply, the uncertain and seasonal equilibrium between the two, and the role of immigration in addressing the farm labor problem. It concludes with a view toward the future, discussing how agriculture must adjust to a new farm labor equilibrium as people move off the farm and the global demand for food continues to rise.

The Farm Labor Problem. https://doi.org/10.1016/B978-0-12-816409-9.00001-X

THE PROBLEM OF FARM LABOR DEMAND

The demand for farm labor derives from farmers' production decisions, which we will learn about in detail in Chapter 2. Agriculture differs from other industries in intrinsic ways that differentiate the timing and magnitude of labor demand. The agricultural production process is biological. It relies heavily on inputs from nature (land and weather). Consequently, agricultural production is highly seasonal. There are long time lags between applying inputs and harvesting outputs. Although farmers can directly control how much they plant, uncertainty surrounds how large the harvestable crop will be and how the harvested crop will be valued on the market. Farming requires land, so agriculture is dispersed over a wide geographic area. Because agricultural production is spread out geographically, involves long time lags, and is highly seasonal and uncertain, timely access to labor, like other inputs, is critical to the success and competitiveness of farm operations.

Agricultural production and access to labor vary substantially around the world. In parts of the United States, large agribusinesses [which the American journalist Carey McWilliams referred to as "factories in the field"] dominate the farm landscape (McWilliams, 2000). Millions of small family farmers dominate production in low-income countries.[1] In most of the world, households, not firms, make most of the agricultural production decisions, and the family provides most or all of the labor needed on the farm. Often, hired labor is an imperfect substitute for family labor, because hired workers might not work as hard, they might not be available for hire near the farm, and poor farmers might lack the cash to hire workers, particularly in the preharvest period prior to receiving payment for the harvest.

Agriculture is marked by a high degree of uncertainty; thus, considerations of risk are a hallmark of agricultural decision-making (Moschini and Hennessy, 2001). There are two broad categories of risk in agriculture: Production risk and marketing risk. On the plus side, nature provides inputs like sunshine and rainfall at no cost. The downside is that farmers cannot predict when the rains will or will not come, whether there will be a long season with no sunshine, or whether a swarm of locusts will devastate the crop. Shocks of nature break the engineering relationship between inputs and outputs. Variables outside farmers' control determine how much of a harvestable crop there will be on the tree or in the field—not only the vagaries of weather, but also risks associated with pests and access to inputs, including labor.

This sort of uncertainty generally does not arise in manufacturing, where engineering relationships govern production processes. Farmers' access to hired workers and other inputs, credit, insurance, and a market for the harvested crop

---

1. There are an estimated 570 million farms in the world. A total of 35% are located in China and 24% in India. Small farms (less than 2 ha) constitute about 12% of farm operations worldwide, and family farms constitute about 75% (Lowder et al., 2016).

impacts how they will respond to seasonality and uncertainty in agricultural production. New evidence suggests that climate change is increasing agricultural production risk as well as impacting agricultural labor markets.[2]

Once the harvestable stock of produce is mature in the fields or on the trees, agricultural production is largely a resource-extraction problem—how to harvest the crop and get it to market in the most economically efficient way. Most of the risk at this stage revolves around the availability of harvest labor and market prices. Seasonal variations in spot-market prices create critical timing windows for agricultural producers. Increasingly, grower-shippers have time-sensitive contractual commitments as preferred suppliers to mass merchandisers (e.g., Walmart and Costco), supermarket chains, and food service industries. This is true in developing as well as high-income countries (Reardon et al., 2003). Failure to harvest a field on time can result in the grower-shipper failing to meet delivery commitments to buyers under Vender Managed Inventory Replacement (VMIR) and other preferred supplier agreements. Some research suggests that, with the consolidation of the retail sector, shippers have less negotiating power and are more fearful of losing accounts if they fail to comply with buyer requests. Increasing trade integration creates price competition and narrows marketing windows, intensifying pressure on farmers. Having access to workers at critical moments in the production process can make the difference between meeting delivery commitments or not.

Labor supply risks are paramount at harvest time. An important potential component of risk is the lack of available labor at the times and places needed to harvest crops. Production risk can result from an insufficient labor supply if fruit spoils on the trees before it can be harvested or if labor shortages prevent farmers from marketing their harvest on time and complying with their contractual obligations higher up on the supply chain.

The Agricultural Production Function

The production function occupies center stage in farmers' decisions, as it does in the microeconomics of all firms. It describes a technological relationship, a recipe to convert inputs into outputs. In most sectors of the economy, the production function represents a known engineering relationship between inputs and outputs, like how many copies of this book can be produced from a given amount of paper, ink, capital (printing machines), labor, and so on. Like a kitchen recipe, a production function can describe a fixed relationship between inputs and outputs. For example, a tomato harvester might pick an average of 30–33 1-pound buckets in an hour, implying around 1.9 min of

---

2. Examples include (Mendelsohn et al., 1994; Schlenker and Roberts, 2009; Deschenes and Greenstone, 2007; Jessoe et al., 2016).

labor per harvested bucket.[3] Most production functions are not linear, though. They reflect decreasing marginal productivity of inputs as well as the possibility that inputs may substitute for one another (unlike flour and salt in a pancake recipe). Increased use of machinery can reduce labor needs. Even in the harvest, diminishing returns eventually set in if too many workers are added to the harvest crew.

The harvest depends on the stock of produce in the orchard or field ready to harvest as well as the inputs applied at harvest time. The harvestable stock, in turn, is uncertain, and it depends on decisions made prior to the harvest. Input-output relationships involved in putting a harvestable crop in the field or on the tree are different from those involved at harvest time. Agricultural production is sequential, with different production functions describing input-output relationships at different stages of the production process. The sequential nature of crop production and risk means that we should not treat farm labor like other production inputs or like labor in other sectors of the economy using a conventional single-stage production model—even though most researchers do just that.

Commercial farmers, like other producers, use the technology at their disposal to combine inputs and produce output, with the objective of maximizing profits. (They may have other objectives as well, but for most purposes, profits are a reasonable focus in agricultural models.) Profit maximization implies employing additional variable inputs, including labor, up to the point where the benefit of adding an additional unit of input (the marginal value product of the input) just equals the input's per-unit price, given the production function. A farmer will not pay employees for an additional day of work unless this results in a sufficient increase in gross revenue to cover the wage bill. If by hiring an additional worker-day of labor the farmer will gain an increase in harvest worth more than the wages paid, she will do it.[4]

Technological Change: Switching to a New Production Function

Over time, producers may invest in new labor-saving technologies to reduce the costs of hiring workers. This implies switching to a new production function. When mechanical harvesters (or more recently, robots) accompany humans in the fields at harvest time, the recipe changes: fewer worker-days are required per pound of harvested crop. The history of agricultural production is largely about the development and adoption of new technologies that change the input-output relationship. Labor-saving technologies, like the cotton gin and combine harvesters, dramatically reduced labor needs in cotton and grain fields.

---

3. These tomato harvest labor figures come from a study entitled "Labor Requirements and Costs for Harvesting Tomatoes," by Zhengfei Guan, Feng Wu, and Steven Sargent, University of Florida; http://gcrec.ifas.ufl.edu/media/gcrecifasufledu/docs/zhengfei/FE-1026—Tomato-Harvest.pdf.
4. One worker-day is a single worker employed for one day, two workers employed for ½ day each, etc.

New IT-intensive technologies use robotics to reduce labor needs for traditionally labor-intensive fruit and vegetable crops. The Green Revolution is another example of technology change. New high-yielding seeds dramatically increased yields the world's major food crops, but Green Revolution technologies are land saving, not labor saving: They enable farmers to produce a larger output on the same amount of land. In most cases, they require additional labor to apply fertilizer and other inputs as well as to bring in a larger harvest.

Adopting a new technology is costly. Farmers will only invest in a new technology if it passes the cost-benefit test (Chapter 9). The discounted benefits (cost savings) from the investment over time must be greater than the cost of the initial investment in order for a technology to be economically feasible. The cost savings depend on the nature of the technology and also on input costs. If the cost of an input (say, farmworker wages) is increasing over time, a labor-saving technology becomes more attractive. If, on the other hand, land is scarce, land rents are rising, and labor is abundant, a land-saving technology is more likely to pass the cost-benefit test.

Farmers are like any commercial producer in that one of their goals is to maximize profits. However, the differences between agriculture and other production sectors have profound implications when it comes to determining the optimal amount and timing of labor inputs with a given technology, or deciding whether it is optimal to switch to a new technology.

Production Risk

The agricultural production process is biological and filled with uncertainty. Thus, agricultural economists pay a great deal of attention to incorporating risk into the crop production function. In agriculture, the production function is random or stochastic [from the Greek word στόχος (stóchos), meaning "guess" or "target"]. We will learn about stochastic production analysis and what it means for labor demand in later chapters, but for now it is important to bear in mind this crucial difference between agriculture and other sectors. The amount of inputs farmers demand with a given technology depends on yield and price uncertainty.

The choice of technology also depends on risk. Often, there is a tradeoff between risk and expected returns from new technologies. A high-yielding seed might require greater investment in fertilizer and other inputs—not to mention labor to apply the inputs—in return for a yield that is higher on average but more variable than yields from conventional seeds. Yield risk is particularly a concern on the marginal, rain-fed lands most of the world's farmers sow. If the rains do not come (or too much rain comes, or wind, or too much heat or cold), farmers risk losing the crop as well as the money they spend on inputs ahead of the harvest. It is not surprising, then, that marginal lands are where one is *least* likely to find high-yielding crop varieties (HYVs) around the world.

New technologies can also reduce risk. Agricultural research currently is underway to develop seed varieties that are robust to drought and other weather shocks, without necessarily increasing expected yield in normal years. In effect, these new seeds come with their own built-in insurance policy.[5] As labor becomes scarcer and its availability becomes less certain, labor-saving crop technologies can reduce production risks while making it possible to harvest the same crop with fewer (and no doubt more expensive) workers. A farmer might be willing to spend more on a labor-saving technology than the wage savings one expects to gain from it, provided that the technology makes one less vulnerable to unexpected labor shortages at harvest time. The difference between the adoption cost and expected wage savings could be justified as a "harvest labor insurance premium."

Timing and Seasonality

Farm labor employment is highly seasonal. A 1938 study of California farm labor noted extreme fluctuations in seasonal labor demand (Barry, 2007):

> *"During the peak season of September of the year 1935...there were demands for more than 198,000 workers, while in the slack season of the same year, during December and January, but about 47,000 were required, leaving over 150,000 unemployed."*

More recent data from the US Bureau of Labor Statistics paint a similar picture. Fig. 1.1 shows the seasonal ups and downs of the two components of the US farm labor demand for fruit, vegetable, and horticultural production between 2001 and 2009: direct hiring by farmers and hiring by farm labor contractors (FLCs), who supply workers to farms. For both, the peak-to-trough ratios appear remarkably stable from one year to the next—and they are big.

Agriculture involves at least two production stages. In the preharvest season, the farmer's objective is to efficiently create a stock of harvestable fruit on the tree (or vine, or field). In the harvest period, his objective is to turn the harvestable stock into a marketable product, picked and packaged. There is no reason to think that the demand for labor looks at all the same in these two periods.

Take a grape farm, for example. It employs a few workers year round to irrigate, weed, prune, and perform other tasks required for the vine to produce fruit. The harvest season, on the other hand, is short and intense. The Fresno, California, raisin grape harvest used to be the most labor-intensive activity in US agriculture, employing 40,000 to 50,000 workers for only a few weeks each fall to pick the grapes and spread them out on paper trays to dry in the sun. Today, many farms have switched to dry-on-the-vine technology, in which machines shake the dried raisins into bins. This reduces the need for farm workers during harvest dramatically, almost eliminating harvest labor risks.

A multistage production process means that agricultural labor demand tends to be highly seasonal. Combine this with risk and we can begin to understand

---

5. For example, see Lybbert (2006).

United States

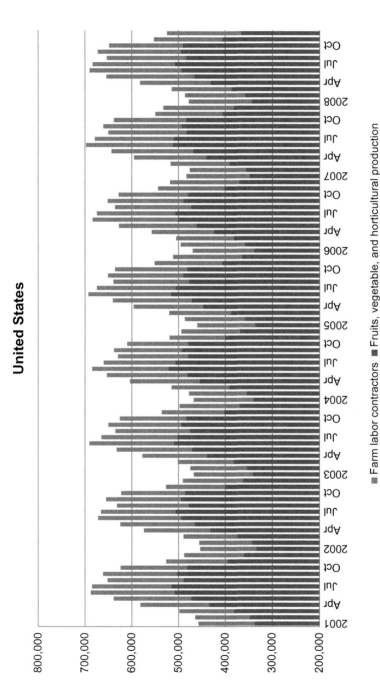

■ Farm labor contractors ■ Fruits, vegetable, and horticultural production

FIG. 1.1 Employment by FVH producers and farm labor contractors in the United States is highly seasonal. (*Data from Bureau of Labor Statistics Quarterly Census of Employment and Wages (http://www.bls.gov/cew/data.htm).*)

why agriculture and farm labor markets are so different. Agricultural production *at each stage of the production process* is stochastic. Nature's surprises early in the growing season can create large swings in labor demand come harvest time.

Whenever there are time lags between input use and harvest, producers have to find ways to finance their inputs. They also have to find ways to put food on the family table if the crop fails. Credit is vital to most farmers in high-income countries; farmers can take out loans to purchase inputs, and then repay the loans after harvest. These inputs include labor. Workers might be willing to work on your farm, but only if you have the cash before harvest to pay them.

Crop insurance is a different matter. Most farmers who grow labor-intensive fruit, vegetable, and horticultural (FVH) crops do not have an insurance policy that will pay out if the crop fails. Grain farmers in the United States and Europe do have access to crop insurance, but only because of generous government programs.[6]

Credit markets are not available to most farmers in poor countries, and formal crop insurance is virtually nonexistent outside of a few development projects.[7] Yet poor farmers in developing countries, like farmers in rich countries, take on a heavy risk burden when they buy inputs and hire workers prior to harvest, knowing that the crop might fail or the price of the harvested crop might tumble.

Space

Farming requires land, so agriculture is spread over wide geographic areas. Contrast this with, say, information and technology industries in California's Silicon Valley or Bangalore, India, which reap *economies of agglomeration* by locating close to one another so that they can share ideas and build on one another's innovations. Because farms are spread out, timely access to diverse markets is crucial. This includes markets for the output farmers produce, but it also includes markets for inputs—including labor.

Workers have to be available to farms in the right numbers and at just the right times, or else crop losses are likely. A farmer might do everything right in the preharvest periods, growing conditions might be ideal, and there might be a bumper crop in the field, but if workers are not available to pick the crop at the critical moment, the crop can rot before it is harvested. Even if labor shortages leave, say, the last 15% of the crop unpicked, that might represent the farmer's profit margin.

---

6. You can learn about the Crop Insurance Program at the U.S. Department of Agriculture (USDA) website: http://www.rma.usda.gov/aboutrma/what/history.html. The European Union has a less comprehensive crop insurance program; e.g., see (Santeramo and Ford Ramsey, 2017).

7. An introduction to credit and insurance in developing countries appears in Taylor and Lybbert (2015).

On a small family farm, the farmer's family is likely to supply most or all of the labor to cultivate and harvest crops. On many larger farms, including commercial farms throughout the developed world, family labor is largely irrelevant. A large number of hired farm workers are needed for a short period of time to bring in the harvest.

How do farmers find so many workers on short notice? In villages around the world, information flows through networks of contacts between farmers and workers and among workers. Sometimes landlords, worried about having sufficient workers for harvest, enter into complex "labor tying" arrangements in which they provide people with various benefits in return for a promise to work on the landlord's farm at times of peak labor demand. The benefits might include one or a combination of employment during the off season (when people have a hard time finding work), access to plots of land to farm, credit, or inputs like seed and fertilizer.[8]

In John Steinbeck's story of the great Dust Bowl migration, *The Grapes of Wrath*, desperate workers showed up at California farms begging for work at any wage. Many farmers rely on the same workers returning from one year to the next to harvest crops. Some encourage this by providing workers with extra hours of work, higher wages, and nonwage benefits; building personal relationships with individual workers; or even by providing support to the communities from which these workers come (in the US, some come from as far away as villages in southern Mexico). The classic 1960 CBS News documentary "Harvest of Shame" followed migrant workers making their annual trek harvesting beans and other crops up and down the US Atlantic Seaboard—often with children in tow.

Women play a key role in filling the harvest labor vacuum in many parts of the world. In Chile, many women enter the workforce to harvest fresh fruits and vegetables in the peak season, when wages increase, then return to their households after the harvests are in. In Morocco, women are far and away the main source of labor to harvest and process the crocus flower stamen from which saffron is produced. In Burkina Faso and many other African countries, women manage their own small plots of land and supply most of the labor—often with little or no help from men. Although tea harvesting machines exist (including a Japanese version with advanced laser technology), most tea harvesting or "plucking" is performed by women. A 2011 United Nations study found that women comprised about 43% of the agricultural workforce globally and more than 50% in some African countries (Jarvis and Vera-Toscano, 2004; Filipski et al., 2017; Udry, 1996; SOFA Team and Doss, 2011).

Today in California, FLCs match thousands of farm workers with jobs on many individual farms. Instead of hiring workers directly, a farmer can enter into a contract with an FLC to, say, harvest 20 acres of oranges. The FLCs' comparative advantage is their network of contacts with farm workers mobilized on

---

8. A classic work in this area is Bardhan (1984).

short notice and their intensive use of day labor markets, where workers gather in the hope of getting a seat on a contractor's bus.

Inequality and Concentration

Some sectors, like energy, steel, or automobiles, are dominated by a few large producers. Agricultural production is different. It is by far the largest production sector in the world in terms of numbers of producers. Of the world's 570 million farms, more than 475 million are less than 2 ha in size (Lowder et al., 2014). Thus, the vast majority of the world's farms operate only a small share of the world's farmland. The world's farmers have unequal access to resources, from large agribusinesses with first-world infrastructure and market access to small-scale farms producing for farmers' markets to hundreds of millions of poor family farmers struggling for subsistence in less-developed countries.

To understand the supply and demand for farm labor, we have to understand the behavior of thousands, millions, or (in China and India) hundreds of millions of heterogeneous actors. Influencing or simply understanding agricultural outcomes requires having good economic models of diverse agricultural producers and how they are likely to respond to different kinds of market, environmental, or policy shocks. To complicate matters, as we shall see, what is beneficial for large farmers may be detrimental for small farmers.

THE PROBLEM OF FARM LABOR SUPPLY

The farm labor supply comes from households, which decide how much of their time to supply to farm and other work activities in order to maximize their welfare, or utility (Chapter 3). In poor countries, agriculture is large relative to other sectors in the economy, and agricultural labor is in abundant supply. Most countries start out rural and agricultural. In the early stages of economic development, the farm labor supply is elastic (Lewis, 1954). That is, workers are available to satisfy farm labor demands even at low wages, and they quickly increase their labor supply when wages increase.

In high-income countries, agriculture is a small sector compared with industry and services in terms of employment. The human capital (skill) requirements for agriculture are minimal. One might expect, then, that the supply of labor to agriculture, like any small sector, is elastic, easily expanding to meet seasonal increases in labor demand. Contrast this with high human capital jobs, like computer programming and health care, for which it might take several years to train new workers when labor demands increase. Human capital investment (in particular, investment in schooling) is a major focus of dynamic labor supply models. It is essential for doctors but not for farm workers.

Although it is possible to train farmworkers relatively quickly, people leave farm work as economies develop and off-farm employment expands. This results in a shrinking pool of domestic farm workers. In economic parlance, the domestic agricultural labor supply shifts inward. People flee the farm. We can imagine an equilibrium model in which agricultural wages increase apace with nonagricultural wages, inducing domestic workers to stay on the farm instead of moving to factory or service jobs. There might have to be an extra wage premium to induce people to take farm jobs if nonfarm jobs bring other nonpecuniary benefits, like being more interesting and less onerous than farm work. Ask yourself: What would it take to get you out into the fields picking crops on a hot summer afternoon? How long would you be willing to continue doing field work?

History shows that rising farm wages do not induce large numbers of domestic workers to do farm work. You will not find US-born unemployed steel workers picking lettuce in California or oranges in Florida. The share of domestic workers in the US hired farm work force has fallen to the point where, by 2006, only 23% of workers (2% in California) were US-born.[9] The rest were immigrants, earning a wage easily eight times what they could hope to earn by working in rural Mexico, which is where most hired farmworkers in the United States come from.

The wage rate in rural Mexico represents the opportunity cost of migrating to US farm jobs.[10] It is the lower bound on the reservation wage, the lowest wage a worker would accept to migrate to the United States instead of working in Mexico. A rural Mexican's reservation wage also includes the economic and psychic costs of migrating. Virtually all new immigrant farmworkers lack legal status, so their reservation wage reflects the risk of being apprehended, jailed, and returned to Mexico.[11]

The fact that most hired farmworkers in the US earn low wages and come from Mexico suggests that the foreign farm labor supply is elastic. An elastic foreign labor supply at a relatively low wage can explain why the farm workforce in the United States and other rich countries consists mostly of immigrants. The quantity of domestic labor supplied to agriculture drops as nonfarm employment expands, and foreign workers come in to fill the void. Any excess demand for farm workers gets filled by foreign workers, and this keeps wages lower than they would be without immigration. In the United States, a growing reliance on foreign agricultural workers and testimonials from farm workers, whose children usually eschew farm work, support Martin's contention that the farm workers of tomorrow are growing up outside the United States (Martin, 2009).

Within farm labor-importing countries, the supply of labor to individual farms and farm regions may be elastic in the short run, in the sense that a farm

---

9. Estimated from the U.S. Department of Labor's NAWS Public Use Data (http://www.doleta.gov/agworker/naws.cfm).
10. The market wage and marginal value product of labor are the same in a "separable" agricultural household model; see Singh, Squire, and Strauss, 1986.
11. The migrant's reservation wage may include other elements, as well: psychic costs of working abroad, cost of living adjustments, etc.

can attract more workers by offering a slightly higher wage if it needs to. There is an important spatial element to this story, though. Unless the total farm labor supply can quickly adjust, farms can attract new workers only at the expense of other farms. As fewer people in rural Mexico enter farm work, farm wages in the United States are rising. But as one observer put it:

> *The one constant is that no matter how much we pay, domestic workers are not applying for these jobs. Raising wages only serves to cannibalize from the existing workforce; it does nothing to add new laborers to the pool.*

(Kitroeff, 2017)

The impacts of an excess demand for farm labor in one region reverberate into other regions. A late harvest in one locality can prevent workers from migrating to another locality. This produces the familiar pattern of localized seasonal farm labor shortages, even if there is overall labor abundance and average farm wages are low.

EQUILIBRIUM IN FARM LABOR MARKETS

Understanding how farm labor markets work requires integrating demand and supply into farm labor market equilibrium models (Chapter 4). We use the plural—"models"—here because there is no single farm labor market anywhere. Instead, there are many localized farm labor markets, often linked together by migration.

In 2007, a *Wall Street Journal* editorial claimed that "farmers nationwide are facing their most serious labor shortage in years." The editorial asserted that "20% of American agricultural products were stranded at the farm gate" in 2006, including a third of North Carolina cucumbers, and it predicted that total crop losses in California would hit 30% in 2007.

These predictions did not materialize. The example of cucumbers illustrates that farm labor shortages tend to be local, not generalized across the nation. US cucumber production rose both in 2006 and 2007 (Farm Labor Shortages, Mechanization, 2007).

This and many other farm labor-shortage anecdotes suggest that, even if the medium-to-long run supply of foreign workers is elastic, fed by new immigration, in the short run farmers compete across regions for a relatively inelastic, or fixed, total supply of agricultural workers.

In an average year, what might be called a "first-order" equilibrium supported at the margin by follow-the-crop migration ensures ample labor at a predictable wage in a given place and season. However, within a network of migration-linked farm labor markets, a stochastic shock to one local market (e.g., a late pear harvest in the Sacramento Valley) reverberates into neighboring labor markets. One needs multiregion models to represent this spatial complexity of farm labor supply and demand.

A number of researchers point out an important caveat to the seasonal equilibrium just described. They note that the process of adjustment in farm labor markets often is sluggish and incomplete. Studies focusing on developing countries argue that wage rigidities force farm labor markets to adjust through changes in unemployment (Bardhan, 1979; Dreze and Mukherjee, 1989). Jarvis and Vera-Toscano (2004), in a study of seasonal farm labor markets in Chile, found that female workers absorbed most of the seasonal labor force adjustments, because they had the lowest reservation wage.

In the United States, the availability of welfare and unemployment insurance puts a lower bound on the reservation wage. Instead of migrating to follow the harvest, many workers can stay put and receive unemployment insurance and welfare support. Thus, farm labor markets are likely to adjust to demand shocks via changes in the demand for welfare and unemployment insurance, rather than through an endogenous wage adjustment that redistributes farm workers across regions. Minimum wage laws reinforce this. A study by three UC Davis researchers found that seasonal work, low earnings, and high unemployment in California's agricultural counties promoted welfare use and limited the potential of local labor markets to absorb ex-welfare recipients. In California's major agricultural counties, when unemployment peaks, welfare use increases (Green et al., 2003).

A CONTINUUM FROM FAMILY FARMING TO AN IMMIGRANT FARM WORKFORCE

Perhaps the most notable difference between agriculture and other economic sectors in poor countries is that most agricultural production is by households. Most farms in the world are family farms that supply their own labor and other inputs to the land and consume part or all of the harvest. Unlike firms that are focused exclusively on production decisions, agricultural households make both production and consumption decisions. This may seem like a technical distinction, but as we see in Chapter 5, it can drastically change the economic analysis of farm labor demand as well as supply.

Around the world, hired labor markets evolve over time in ways that mirror the agricultural transformation. At low levels of economic development, family farming dominates the agricultural landscape. Agricultural households provide most of the agricultural inputs and consume a significant part (in most countries, most) of the output from the farm. Not surprisingly, in predominantly subsistence economies, agricultural labor markets are relatively "thin" and often take the form of informal labor exchanges. In some areas, large plantations create a demand for wage labor that coexists with subsistence production on small farms.

The importance of hired labor expands as commercial production displaces family production. Consider, for instance, the United States, where both family and hired labor have declined over the last half century as a result of mechanization, but the ratio of hired to family labor has increased, from 1:3 in 1950 to around 1:2 today (these averages mask large differences across US regions) (Kandel, 2008). Increasing reliance on hired workers frequently gives rise to internal migration to address labor market disequilibria across time and space. However, a collection of factors—seasonality, uncertainty of farm employment, aversion to follow-the-crop migration, the disagreeableness of working conditions in the field, and expanding income opportunities outside of agriculture—cause the domestic supply of agricultural labor eventually to shift inward and upward. This creates incentives for farmers to seek a less expensive source of labor abroad.

Migration policies evolve in imperfect harmony with farm labor market trends, because they are the outcome of a complex and frequently internally contradictory political process (Chapter 6). Rarely do countries implement policies explicitly to influence internal migration. Nevertheless, at the mid-stage of the agricultural transformation, there tends to be growing concern for the welfare of migrant farm laborers, their families, and the communities in which they live. In some places, most notably California, farm labor organizing efforts escalate (Chapter 7). At this stage, as Emerson (1989) states, "a major emphasis of governmental policy toward migratory farm labor is to shift the migrant out of the migratory stream, and if at all possible, to shift him to the nonfarm sector."[12]

This gives rise to an integration-immigration dilemma: In the absence of labor-saving technological innovations and/or a shift to crops requiring less labor, mobility out of agriculture for some workers implies the rotation of new immigrants into the farm workforce. Martin (2009) describes an agricultural "immigration treadmill" that perpetuates rural poverty by attracting a continual flow of new unskilled immigrants to take the place of workers who move from farm to nonfarm jobs.

In high-income economies, agriculture loses its importance as a generator of employment, while its dependence on foreign workers invariably increases. The political process becomes a battleground in which farm interests engage the interests of other actors, some of whom oppose the use of immigration policies to guarantee an abundant supply of agricultural labor.

If border enforcement restricts immigration, the farm labor supply decreases and wages increase. Meanwhile, the juxtaposition of high farmworker wages in high-income countries with low earnings abroad keep the supply of immigrant labor elastic, willing to cross borders to fill farm jobs. This intensifies pressure at the border and makes immigration policies difficult to enforce. According to conservative estimates, more than 50% of all US hired farmworkers are

---

12. Examples of government programs include the migrant and seasonal farm worker programs launched as part of the War on Poverty in the 1960s, including Migrant Headstart, Migrant Education, Migrant Health, and Job Training for Seasonal and Migrant Workers (Emerson, 1989).

unauthorized immigrants.[13] The unauthorized share of workforces in other high-income countries are not as well documented but certainly large.

Instead of seeking legal access to farmworkers from abroad, farmers could adopt labor-saving technologies, invest in improved labor management practices, or take steps to evade immigration laws. There is evidence that penalties against knowingly hiring unauthorized immigrants, which were included in the United States 1986 Immigration Reform and Control Act (IRCA), accelerated a shift away from direct hiring by farmers in favor of labor intermediaries, who are more difficult for immigration law enforcement authorities to monitor (Taylor and Thilmany, 1993).

THE END OF FARM LABOR ABUNDANCE

A pervasive theme in this chapter is that domestic agricultural workers become increasingly scarce as countries' incomes rise. Both logic and emerging empirical evidence suggest that the same applies to the supply of foreign agricultural labor. Labor-intensive agriculture in high-income countries seeks out new migrant-source areas over time. What happens when foreign agricultural workers move out of farm work? The evolution of agricultural labor markets in farm labor-exporting regions raises questions about the sustainability of a labor-intensive agricultural system dependent on immigrant labor, and indeed of labor-intensive agricultural systems generally (Chapter 8).

Nowhere are agricultural labor markets more integrated across borders than in Mexico and the United States. New research finds that the supply of farm labor from rural Mexico is diminishing over time (Taylor et al., 2012; Charlton and Taylor, 2016). United States and Mexican farmers compete for an ever-smaller number of farm workers from rural Mexico. Migration is network driven; new migrants tend to follow past migrants to the same destinations and jobs (Sana and Massey, 2005; Mora and Taylor, 2006; Pfeiffer et al., 2007). As immigrant workers move out of agriculture and into factory and service jobs, future migrants do, too. The supply of labor to agriculture diminishes.

In Mexico, the regional shift in labor supply to US farms is happening quickly. Mexico has entered into agreements with Guatemala to import Guatemalan farmworkers.[14] Thus, Mexico is in a transitional phase of being both an importer and exporter of farm workers. With a combined population one-third the size of Mexico's and accelerating urbanization, Central America's potential

---

13. U.S. Department of Agriculture, Farm Labor. http://www.ers.usda.gov/topics/farm-economy/farm-labor/background.aspx#legalstatus.

14. México Centro de Estudios Migratorios/Instituto Nacional de Migración, Trabajadores guatemaltecos documentados con la Forma Migratoria de Visitante Agrícola (FMVA) en el estado de Chiapas; http://www.gobernacion.gob.mx/work/models/SEGOB/Resource/2796/1/images/Dossier%206_%20Trabajadores%20guatemaltecos%20documentados%20con%20la%20forma%20migratoria%20de%20visitante%20agr%c3%83%c2%adcola%20(FMVA)%20en%20el%20estado%20de%20Chiapas.pdf

to replace Mexico as a significant source of US farm labor is limited. A 2017 headline in a major Mexican newspaper declared: "In agriculture, Mexico is left without young producers (Georgina Saldierna, 2017)."

In Western Europe, immigrants from African nations work the fields of Spain, France, and Italy; Albanians harvest crops in Greece; Polish workers toil the fields in Germany and the United Kingdom. Regional integration under the European Union (EU), including free labor movement among EU countries under the Schengen Accord, boosted farm labor migration from the former Soviet bloc countries. In time, one would expect that the migration networks currently channeling eastern European workers into agricultural jobs in the EU will become more urbanized, like networks from Mexico to the United States.

Worried about its heavy reliance on foreign agricultural workers, Germany offered farmers generous subsidies to hire German-born farm workers. That experiment failed, though, and the program was abandoned (Txortzis, 2006). Israel offered a different alternative to its farmers. In December 2009 it announced a new public initiative to invest in mechanization of Israeli farming "in an effort to reduce the need for foreign workers (Arutz Sheva, 2009)."

ROBOTS IN THE FIELDS

Countries dependent on imported agricultural labor have two options. The first is to seek out new sources of farmworkers and (through immigration policy) grant farmers access to them. This is not a long-run solution, though, because the same process that shifts workers out of agriculture as economies grow and contentious political processes limit rich countries' ability to find—and import—new sources of cheap farm labor over time.

The second option is to invest in reducing farm labor demands through a combination of technological change, improved labor management practices, and trade. Instead of large crews of low-skilled and poorly paid workers toiling in the fields, this second option implies better-paid, skilled workers accompanying robots in the fields (Chapter 9).

Stepping into a Wellsian time machine and glimpsing the agricultural future in today's farm labor-importing countries like the United States, we are sure to find a combination of higher capital intensity and labor productivity, big labor-saving innovations involving machine learning and robotics, a shift to more imports of crops that are hard to mechanize, and higher farm wages.[15] The farm workforce will change; tekked-up workers will be needed to work in a tekked-up agriculture.

The transition to this new agricultural world is likely to proceed unevenly across countries, commodities, and agricultural tasks. The profitability of labor-saving technological changes and crop mixes at a particular place and

---

15. *The Time Machine* is a science fiction novel by H.G. Wells.

time depends on the availability of low-cost labor. Policies facilitating access to low-wage immigrant labor and developments abroad that stimulate emigration will continue to discourage the adoption of labor-saving technologies. Mechanical harvesters and new IT-assisted labor-saving solutions exist or are under development for most crops, their usage limited mostly by low wages for harvest workers and quality and productivity concerns that can be addressed by future research and development.

New developments in information and technology, machine learning, and artificial intelligence point to a future of robot-assisted agricultural production. Engineering departments at universities and high-tech startup firms in Silicon Valley are beginning to develop labor-saving solutions for a diversity of crops that combine mechanical engineering with artificial intelligence and machine learning. The perishable nature of FVH crops, imperfect substitutability between capital and labor, and consumer demands for high-quality locally grown produce will insure that some production of labor intensive crops persists even in high-wage countries. In time, though, the farm labor-immigration policy connection is bound to weaken, as changes in the availability of low-cost labor, technologies, crop mixes, and trade reduce countries' reliance on imported agricultural workers.

REFERENCES

Bardhan, P.K., 1979. Wages and unemployment in a poor agrarian economy: a theoretical and empirical analysis. J. Politic. Econ. 87 (3), 479–500.

Bardhan, P.K., 1984. Land, Labor, and Rural Poverty. Cambridge University Press, Cambridge.

Barry, 2007. Influence of Employment Agencies on Migratory Farm Labor. p. 4.

Charlton, D., Taylor, J.E., 2016. A declining farm workforce: analysis of panel data from rural Mexico. Am. J. Agric. Econ. 98 (4), 1158–1180.

Deschenes, O., Greenstone, M., 2007. The economic impacts of climate change: evidence from agricultural output and random fluctuations in weather. Am. Econ. Rev. 97 (1), 354–385.

Dreze, J.P., Mukherjee, A., 1989. Labour contracts in rural India: theories and evidence. In: The Balance Between Industry and Agriculture in Economic Development. Palgrave Macmillan, London. pp. 233–265.

Emerson, R.D., 1989. Migratory labor and agriculture. Am. J. Agric. Econ. 71 (3), 617–629.

Farm Labor Shortages, Mechanization, 2007. Rural Migration News. vol. 13. No. 4 (October 2007). Cucumber production Statistics are from USDA, http://www.nass.usda.gov/QuickStats/index2.jsp.

Filipski, M., Aboudrare, A., Lybbert, T.J., Taylor, J.E., 2017. Spice price spikes: simulating impacts of saffron price volatility in a gendered local economy-wide model. World Dev. 91, 84–99.

Green, R., Martin, P., Taylor, J.E., 2003. Welfare reform in agricultural California. Agric. Resour. Econ. Rev. 169–183.

Jarvis, L., Vera-Toscano, E., 2004. Seasonal adjustment in a market for female agricultural workers. Am. J. Agric. Econ. 86 (1), 254–266.

Jessoe, K., Manning, D., Taylor, E.J., 2016. Climate change and labour allocation in rural Mexico: evidence from annual fluctuations in weather. Econ. J. https://doi.org/10.1111/ecoj.12448/fullhttp://onlinelibrary.wiley.com.

Kandel, W., 2008. Profile of Hired Farmworkers, a 2008 Update (Economic Research Report Number 60) [Electronic Version]. Economic Research Service, United States Department of Agriculture, Washington, DC.http://digitalcommons.ilr.cornell.edu/key_workplace/559/.

Kitroeff, N., 2017. How this garlic farm went from a labor shortage to over 150 people on its applicant waitlist. Los Angeles Times. February 9, 2017, http://www.latimes.com/business/la-fi-garlic-labor-shortage-20170207-story.html.

Lewis, W.A., 1954. Economic development with unlimited supplies of labour. Manch. Sch. 22 (2), 139–191.

Lowder, S.K., Skoet, J., Singh, S., 2014. What do we really know about the number and distribution of farms and family farms worldwide? Background paper for The State of Food and Agriculture 2014, https://ilcasia.wordpress.com/2014/07/31/what-do-we-really-know-about-the-number-and-distribution-of-farms-and-family-farms-in-the-world/.

Lowder, S.K., Skoet, J., Raney, T., 2016. The number, size, and distribution of farms, smallholder farms, and family farms worldwide. World Dev. 87, 16–29.

Lybbert, T.J., 2006. Indian farmers' valuation of yield distributions: will poor farmers value 'pro-poor' seeds? Food Policy 31 (5), 415–441.

Martin, P.L., 2009. Importing Poverty?: Immigration and the Changing Face of Rural America. Yale University Press.

McWilliams, C., 2000. Factories in the Field: The Story of Migratory Farm Labor in California. University of California Press, Oakland.

Mendelsohn, R., Nordhaus, W., Shaw, D., 1994. The impact of global warming on agriculture: a Ricardian analysis. Am. Econ. Rev. 84 (4), 753–771.

Mora, J., Taylor, J.E., 2006. Determinants of migration, destination, and sector choice: disentangling individual, household, and community effects. In: International Migration, Remittances, and the Brain Drain. pp. 21–52.

Moschini, G., Hennessy, D.A., 2001. Uncertainty, risk aversion, and risk management for agricultural producers. Handb. Agric. Econ. 1, 87–153.

Pfeiffer, L., Richter, S., Fletcher, P., Taylor, J.E., 2007. Gender in economic research on international migration and its impacts: a critical review. In: The International Migration of Women. pp. 11–51.

Reardon, T., Timmer, C.P., Barrett, C.B., Berdegué, J., 2003. The rise of supermarkets in Africa, Asia, and Latin America. Am. J. Agric. Econ. 85 (5), 1140–1146.

Saldierna, G., 2017. En Materia Agrícola, México Se Queda Sin Productores Jóvenes. La Jornada. October 15, 2017, http://www.jornada.unam.mx/2017/10/15/sociedad/033n1soc.

Sana, M., Massey, D.S., 2005. Household composition, family migration, and community context: migrant remittances in four countries. Soc. Sci. Q. 86 (2), 509–528.

Santeramo, F.G., Ford Ramsey, A., 2017. Crop insurance in the EU: Lessons and caution from the US. EuroChoices. https://doi.org/10.1111/1746-692X.12154/fullhttp://onlinelibrary.wiley.com.

Schlenker, W., Roberts, M., 2009. Nonlinear temperature effects indicate severe damages to US crop yields under climate change. Proc. Natl. Acad. Sci. 106 (37), 15594–15598.

Sheva, A., 2009. Israel National News Briefs. In: NIS 625 M to be Invested in Mechanization of Agriculture. http://www.israelnationalnews.com/News/Flash.aspx/177179.

SOFA Team, Doss, C., 2011. The Role of Women in Agriculture. ESA Working Paper No. 11-02, United Nations Food and Agricultural Organization.http://www.fao.org/docrep/013/am307e/am307e00.pdf.

Taylor, J.E., Lybbert, T.J., 2015. Essentials of Development Economics. University of California Press, Oakland, CA (Chapter 12).

Taylor, J.E., Thilmany, D., 1993. Worker turnover, farm labor contractors, and IRCA's impact on the California farm labor market. Am. J. Agric. Econ. 75 (2), 350–360.

Taylor, J.E., Charlton, D., Yúnez-Naude, A., 2012. The end of farm labor abundance. Appl. Econ. Perspect. Policy 34 (4), 587–598.

Timmer, C.P., 1988. The agricultural transformation. Handb. Dev. Econ. 1, 275–331.

Txortzis, A., 2006. When Germans join migrant field hands, the harvest suffers. Christ. Sci. Monit. https://www.csmonitor.com/2006/0524/p01s04-woeu.html.

Udry, C., 1996. Gender, agricultural production, and the theory of the household. J. Polit. Econ. 104 (5), 1010–1046.

Chapter 2

Agricultural Labor Demand

In agriculture, uncertainty is certain.[1]

Glen Cope, fourth-generation cattle rancher from Missouri

Agriculture is different from most other sectors of the economy, because many months elapse from planting to harvest, and farms receive many of their inputs from nature without cost. The dilemma is that nature's supplies of these inputs are uncertain. Thus, so is agricultural labor demand. This chapter discusses the determinants of labor demand in agriculture; why farm labor demand is often so seasonal and volatile; and the implications for agricultural producers, workers, and policymakers.

Commercial farmers, like other producers, use technology to combine inputs to produce an output, with the objective of maximizing profits. Farmers may have other objectives as well, especially if they are part of agricultural households that also consume what they produce (Chapter 5), but it is reasonable to assume that profit maximization is their main objective. At any point in time, profit maximization occurs at the point where the output gain from employing an additional unit of an input just equals the input's price. A farmer will not pay for an additional day of harvest labor unless the labor produces additional value at least as large as the daily wage. If an additional day of labor brings in additional harvest that is valued higher than the day's wage, the farmer will hire the additional day of labor. Over time, producers may invest in new technologies that reduce the number of workers required to harvest the same amount of crop. Farmers will invest in these technologies if the savings from hiring fewer workers exceeds the costs of purchasing the new technologies.

In this way, farmers are like any commercial producer. They choose technologies and decide how many workers to hire and what quantities of other variable inputs to use in order to maximize profits. However, agriculture is different from other sectors in ways that have profound implications for labor demand. Chapter 1 outlined some of the key differences between agriculture

---

1. https://www.drovers.com/article/agriculture-uncertainty-certain (Accessed 10/12/2018).

The Farm Labor Problem. https://doi.org/10.1016/B978-0-12-816409-9.00002-1

and other sectors. When it comes to understanding farm labor demand, two of the biggest differences are that agricultural production is sequential and it entails significant production risk.

Considerable time passes between the moment farmers make their first production decisions and when the crop is harvested. Agriculture really involves at least two production activities. In the preharvest season, the farmer's goal is to create a stock of harvestable produce in the field (or orchard, or vineyard). In the harvest period, her goal is to turn that stock into a marketable product, picked and packaged. There is no reason to think that the demand for labor looks at all the same in these two periods. For example, a grape farm employs a few workers year-round to irrigate, weed, prune, and perform other tasks required for the vines to produce fruit. However, the harvest season is short and intense. A multistage production process means that agricultural labor demand tends to be highly seasonal.

Agriculture also differs from other industries in that, while many of its inputs are provided by nature without cost (sunlight, rainwater, warmth), the provision of these inputs is uncertain. Relatively small variations in weather early in the growing season can produce large swings in labor demand at harvest time. Uncertain states of nature imply that the agricultural labor demand is stochastic, or variable.

MODELING AGRICULTURAL LABOR DEMAND

Fig. 2.1 illustrates a production model with two inputs: capital and labor. A mathematical presentation of the same appears in Appendix A. The model assumes that capital is fixed, that is, the farmer cannot buy or sell capital during

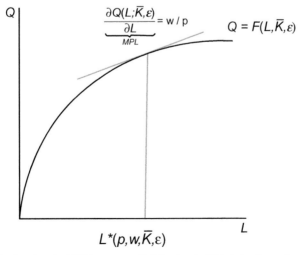

FIG. 2.1   The farm's output (Q) increases with labor inputs (L) but at a decreasing rate.

the growing season. For instance, a peach farmer cannot immediately plant more trees in response to favorable weather conditions, since orchards take years to grow. Nor can a field-crop farmer immediately plant additional acres to get a bigger harvest, since it takes several months for seeds to reach the harvest stage.

In this simple illustration, labor is the only variable input the farmer can adjust in the short run. This production model displays two critical characteristics of production functions: output increases as the input of labor increases, and output rises by a smaller amount with each additional unit of labor employed. Consequently, the production curve is very steep when little labor is employed (when few workers are employed, each additional unit of makes a large contribution to the total output). Then the curve flattens out. The more labor applied to a given plot of land, the less an additional unit of labor contributes to increased production.

The output that an additional unit of labor can produce is called the marginal product of labor (*MPL*). The value of that output (the market price of the output times the *MPL*) is the marginal value product of labor (*MVPL*). The *MVPL* is a very important concept in production economics due to its role in profit maximization. Logically, an employer will not hire an additional worker unless the value the worker produces exceeds the wage the employer must pay. The employer will hire an additional worker if the *MVPL* is greater than the wage. Farmers maximize profits, then, by hiring labor up to the point where the *MVPL* equals the wage.

We can illustrate this optimality condition graphically as point A in Fig. 2.2, where the market-given wage line intersects the *MVPL* curve. The *MVPL* curve

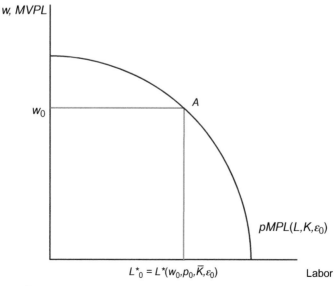

FIG. 2.2 Profits are maximized at the point where the marginal value product of labor equals the wage.

slopes downward because each additional unit of labor produces a little less value than the previous unit of labor. This is a typical property of production functions.

AN UNCERTAIN RELATIONSHIP

In most nonagricultural production activities, the production function represents an engineering relationship. If you purchase and apply the inputs, the function will give the respective output. There is always some error in production calculations, but in, say, an Intel processor plant, there is little doubt about how many processors of a given kind will result from a given combination of labor, capital, and other inputs (e.g., silicon).

This is not the case for agriculture. Most input decisions must be made long before the harvest—before farmers can see how much fruit is on the vine. Farmers continually walk their fields to see how their crops are doing. They know what to look for at every stage: sprouting; plant height, density, and color; blossoms; fruit sets; fruit growth and other properties; and finally, a stock of fruit on the tree or in the field ready to pick. If the state of nature changes, the predicted crop can change, sometimes dramatically. If Q represents the average or expected output per acre, the actual or realized output can be represented as εQ, where ε represents the effect of weather shocks ($\varepsilon = 1$ in a normal year; it is >1 in an above-average year and <1 in a below-average year). An adverse shock will cause the labor demand to fall, as illustrated in Fig. 2.3. The negative weather shock reduces the MPL, and thus the $MVPL$.

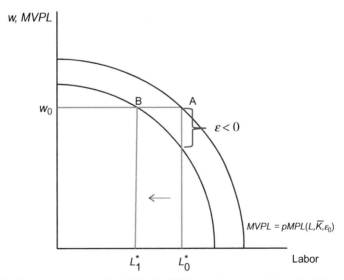

FIG. 2.3   The negative weather shock shifts the $MVPL$ curve inward, and the optimal labor demand falls from L_0^* to L_1^*.

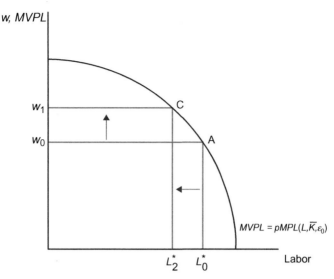

FIG. 2.4 An increase in the wage from w_0 to w_1 results in a decrease in the farm labor demand from L_0^* to L_2^*.

Because $MVPL = pMPL$, a negative output-price shock affects labor demand and output in the same way as a negative weather shock, and Fig. 2.3 can be used to illustrate both.

The impact of a wage shock—for example, an increase in the wage due to a decline in the farm labor supply—is illustrated by a movement along the $MVPL$ curve, not a shift in the curve. Fig. 2.4 shows the impact of an increase in the wage from w_0 to w_1. The labor demand falls from L_0^* to L_2^*. Because output increases with labor input, crop production falls with a positive wage shock.

AN EXAMPLE: LETTUCE LABOR DEMAND ON CALIFORNIA'S CENTRAL COAST

California's central coast—Monterey, San Benito, and Santa Cruz counties—is Steinbeck country. John Steinbeck immortalized this region and its people's struggles to survive off farm wages in a novel about a strike by agricultural workers organized by two communists, entitled *In Dubious Battle* (1936). Agricultural producers were furious. Already confronted by a growing labor movement, they feared that Steinbeck's novel would stir up trouble and add to the uncertainties they already faced.

Steinbeck country is also labor-intensive agricultural country. One of the Central Coast's most labor-intensive crops is lettuce. Lettuce production has evolved substantially since Steinbeck's time, but it is still labor intensive.

At harvest time, a large platform carrying lettuce packers inches its way through the fields, followed by workers who stoop and cut the heads of lettuce, then toss the cut heads to workers on the platform who pack the lettuce into boxes.

We can use a theoretical model of labor demand, together with data on lettuce production and costs, to see how labor-intensive lettuce production is. Appendix A takes us through all the steps to do this. It begins by assuming that the technology that turns intermediate inputs (seeds, fertilizer, etc.) into lettuce output can be described by a Cobb-Douglas production function with labor and capital factor inputs. Cobb-Douglas production functions allow for nonlinearity in the relationships between the factor inputs and output and are relatively easy to estimate with field data. The general form of this production function is:

$$Q = AL^{\alpha}\overline{K}^{1-\alpha}$$

where Q is lettuce output in cartons per acre, $0 < \alpha < 1$, A is a positive shift parameter, L is labor input per acre in hours (h), and K is capital. The bar over K reminds us that capital is fixed in the short run. Appendix A shows that, assuming profit maximization, α is the share of labor in total value-added per acre, and the profit-maximizing per-acre labor demand is:

$$L = p\alpha Q/w$$

where p is the price of a carton of lettuce at the farm gate and w is the hourly farmworker wage. This labor demand function makes intuitive sense. It tells us that the demand for labor increases with output price p, labor productivity α, and per-acre yield Q, and it decreases with the wage, w. As explained in Appendix A, this labor demand function takes into account that the farm uses intermediate inputs in addition to the factor inputs that are arguments in the production function.

We need some data if we wish to apply this labor-demand function to the real world. The University of California Cooperative Extension assembled detailed data on the costs and returns for wrapped iceberg lettuce produced in California's Central Coast region (the data are described in the Appendix). Using these data and following the steps in the Appendix, we obtain a labor demand function of

$$L = 0.27\,pQ/w$$

We can use this equation to predict the per-acre quantity of labor demanded at any output price, yield, and wage. At the midpoint or average lettuce price ($11 per box), yield (950 boxes per acre), and wage ($10.56 per hour in 2016), this equation gives a labor demand of 269.76 h/acre, as pictured in Fig. 2.5.

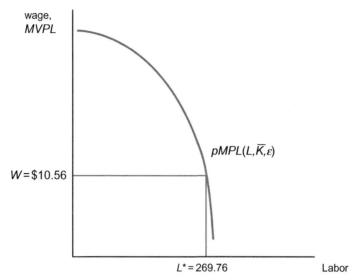

FIG. 2.5   At a wage of $10.56/h, Central Coast lettuce farmers demand an estimated 269.76 hours of labor per acre. *(Based on model parameterized from UC Davis Cost Studies data available at: http://coststudies.ucdavis.edu (evaluated at midpoint yield and price).)*

RETHINKING THE PRODUCTION FUNCTION

In reality, our simple production model might not depict the relationship between inputs and outputs in agriculture as well as it could. Agricultural production does not happen in a single stage, as our model assumes. It involves at least two distinct stages, and probably several. At each stage, inputs are combined to produce a specific outcome, culminating in a harvestable crop. In the preharvest stages of tree-fruit production, for example, the goals are to create blossoms on the trees, good fruit sets, and ultimately a harvestable crop. The final stage is the harvest, itself, which essentially is a resource-extraction problem: farmers use labor and capital to turn a stock of fruit on the vine, tree, or field into a marketable output, like bushels of apples or cartons of lettuce.

The technology for combining labor with capital and other inputs at each stage can be described by a stage-specific production function. At each stage there is an output that may be hard for an economic modeler to ascertain, but the farmer understands. It is whatever the farmer looks for when she walks her fields: young, healthy sprouts, blossoming plants, fruit sets, a crop ready to harvest. The output from each stage is an input into the next stage.

Labor demands may look entirely different from one stage to another. To illustrate this, look again at the data on lettuce production in the Appendix and compare the labor demands before and at harvest. In the preharvest stage, wages constitute 8.5% of total cost ($350/$4102/acre). At harvest, wages are 46% ($2520/$5480).

The uncertain role of nature presents a challenge to economic modelers. If the rains do not come or if they come at the wrong time, the harvestable crop will be small, and fewer workers will be needed to pick the crop. There will be a low output and also a low labor input. It might *look* like the low labor input caused the low output, but really it was the reverse. Low output caused a low demand for labor.

How do we reflect these peculiar features of agriculture in our model, so that we can understand the farm labor demand better? It does not seem right to use the same approach for agriculture as for a computer chip factory. Getting the agricultural production model right is important for studying farm labor demand, because farmers have to find ways to satisfy vastly different labor demands at different stages of production. Exogenous shocks to crop production—good or bad—at any stage alter the labor demand in subsequent stages, particularly at harvest time.

A Two-Stage Crop Production Function

We can analyze the demand for agricultural labor using the two-stage model derived formally in Appendix B. In the first stage, farmers maximize *expected* profits, given a set of assumptions about the weather and market prices during the harvest season, by employing preharvest labor to produce a crop in the ground or on the tree. During this preharvest period, weather conditions are realized and market prices for agricultural goods are settled.

The second stage of production is the harvest. Adverse weather conditions during the preharvest stage imply that the amount of harvestable crop is smaller than originally expected. Better-than-average weather implies the opposite. Lower than expected prices imply that the *MVPL* for any given level of labor is lower than expected, but high prices imply a high *MVPL*.

At the harvest stage, the farmer knows with certainty several factors that she did not know at the beginning of the crop season. Given the realized harvestable crop, wages, and output prices, the farmer determines how much labor to employ at harvest to maximize profits. If weather conditions or prices differ from expectations, then actual labor demand in the second stage of production will differ from the expected demand for harvest labor at the beginning of Stage 1.

The preharvest labor demand L_P^* (where the subscript p refers to preharvest) is based on the assumption of average weather and other shocks. Farmers might revise their assumptions as they learn more about the uncertainties of crop production, or if they perceive changes in these uncertainties over time. Climate change affects farmers' expectations around the world in ways that we do not fully understand yet.

Different farmers might respond to the same perceived uncertainties differently, depending upon their willingness to take on risks (or conversely, their preferences to avoid risk—i.e., their risk aversion). A risk-averse producer weighs the impacts of his decisions on expected profits against the impacts on profit risk. We could consider farmers' risk aversion explicitly, but it would complicate our model. We can say something about how risk aversion would affect the demand for labor and other inputs, though: A risk-averse farmer will tend to act conservatively. That is, she will invest fewer resources in the preharvest season, being worried about potential losses, than a farmer who is less risk averse. Consequently, risk-averse farmers will have a smaller expected stock of harvestable crop than would be the case if risk did not influence farmers' decisions. If crop insurance is available, it could make producers more efficient by reducing risks associated with crop loss.

Pessimistic views on weather and other shocks, and high levels of farmer risk aversion, result in a lower demand for preharvest labor. Even if a negative shock does not materialize and the crop turns out great, fewer inputs in the preharvest period will produce a smaller harvestable crop and less demand for harvest labor. In this way, the preharvest and harvest demands for labor are intricately connected with one another. If a negative shock hits during the preharvest period, the harvestable stock will go down further, and so will the harvest demand for labor.

These problems are magnified in poor countries, where.

- There are no public safety nets to put food on the table if crops fail;
- Farmers do not have the cash or credit to buy inputs;
- There is no crop insurance; and
- Poor roads, communications, and other problems cut farmers off from outside markets for outputs and inputs.

In such environments, few farmers hire labor or come close to using the optimal input mix to produce their crops. As a result, their potential harvests are far lower per acre than in high-income countries. This is one of the great challenges in development economics.

VOLATILITY IN FARM LABOR DEMAND

We have seen that adverse shocks, along with farmers' expectations about these shocks and their reactions to them, affect the demand for labor both before and at harvest in complex ways. This is an aspect in which agriculture is fundamentally different than other sectors of the economy, in which the relationship between inputs and output is a known, engineering relationship, as in a recipe.

How much do weather and other shocks affect farm labor demands? We can use our lettuce labor demand model to explore this. We can also use more complex, multistage production functions to see how weather shocks early in the season have magnified impacts on labor demand at harvest time.

Impacts of Yield and Wage Shocks on Lettuce Labor Demand

Let us look again at our lettuce labor-demand equation:

$$L = 0.27 pQ/w$$

The University of California Cooperative Extension's Cost and Returns Studies give information on the range of per-acre lettuce yields as well as on the range of prices per carton of lettuce that farmers see from year to year. These appear in the bottom of Appendix A, Table 2.A1. Lettuce yields range from 800 to 1100 cartons per acre. The price per carton ranges from $9 to $13. By plugging these different yields and prices into the labor demand equation, we can simulate the changes in lettuce labor demand that result from yield and price variability.

The ranges of yields and prices in the cost report are shown in the rows and columns, respectively, of Table 2.1. The cells in this table show the per-acre labor demands (in hours) given by our labor demand equation under different yield/price scenarios.

The labor demands range from a low of 186 hours, under a low-price, low-yield scenario, to twice that much—369 hours—under a high-price, high-yield scenario. Note that the labor demands in Table 2.1 are not all equally likely to occur. When yields are low, prices tend to rise, and bumper crops tend to drive prices down. Nevertheless, it is clear from this table that labor demands in this crop can vary considerably, depending on states of nature and the market.

Now imagine many crops, each with volatile labor demands, scattered across regions and seasons. That is what the demand side of the farm labor market looks like.

Weather Shocks and Labor Demand in a Multistage Production Function

We can use more advanced simulation methods to explore how positive and negative shocks before the harvest affect the demand for preharvest as well

TABLE 2.1 Per-Acre Labor Demands for Central Coast Lettuce, Under Alternative Price-Yield Scenarios

Yield (Cartons/Acre)	Price ($/Carton)		
	9	11	13
800	185.86	227.17	268.47
950	220.71	269.76	318.81
1100	255.56	312.35	369.15

as harvest labor for all crops in California. Appendix B shows how to derive the demand for preharvest and harvest labor in a two-stage production function. Taylor (2010) divided the preharvest stage into multiple periods. Here is how he did the simulation:

- The crop year was divided into 6 periods (months).
- Each period has its own production function, like the ones we learned about above. They combine inputs at each stage (month) to produce an intermediate output, for example, seedlings at the end of the first month, larger plants the following month, blooms and fruits in the following months, and finally, a harvestable crop in month 5, which is harvested in month 6. A Cobb-Douglas production function with weather shocks, like the one in Appendix B, was used to represent each stage of production. The weather shock variable, ε, enters into each period's production function multiplicatively, like this:

$$Q = \varepsilon A L^{\alpha} K^{1-\alpha}$$

  Assuming that ε multiplies the production function in each period is the simplest way to bring weather shocks into the model. We can think of $\varepsilon = 1$ representing a normal weather period, $\varepsilon > 1$ an above-average period, and $\varepsilon < 1$ a below-average period.
- We have to make an assumption about how weather shocks are distributed. The easiest way to do this is to assume that ε has a normal or "bell-curve" distribution with a mean of 1 and a given standard deviation, which we set equal to 20% of output in each risky period. We used a random number generator to pick ε in each period.
- There is no risk in the harvest period: what you see on the vine is what you can harvest. We let farmers adjust their expectations. When a negative shock hits, the farmer adjusts his expectations downward in the next period, and he does the opposite when a positive shock (unusually good weather) hits. Making many random draws from a weather distribution then simulating the impacts is an example of what is called a Monte Carlo simulation.
- We need parameters for our model—specifically, a shift parameter and exponents for the Cobb-Douglas production function in each period, as in our lettuce example. The study used real data on farm labor demand and assumed a 30% share of labor in total crop value added, which is about what the average labor share is for California specialty crops. (Remember that in a Cobb-Douglas production function, the exponent on labor is the share of labor in total value-added, assuming profit maximization.)

In this simulation, like in real life, a negative shock in one period has two kinds of effects on future labor demands. First, it reduces output and labor demands directly. Second, it makes the farmer rethink his production decisions, cutting back on future inputs to reduce his exposure to the shock. For example, if the expected rains do not come in the first period, farmers cut back on the inputs

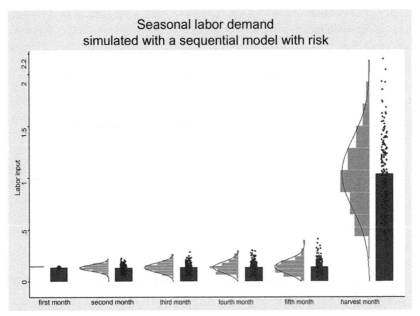

FIG. 2.6 Unexpected shocks to labor demand early in the crop season have impacts that are magnified at harvest time. *(From Taylor, J.E., 2010. Agricultural labor and migration policy. Ann. Rev. Resour. Econ. 2, 369–393, created with the assistance of Mateusz Filipski.)*

they use in later periods, thinking the crop will be poor regardless of subsequent weather conditions. That is, we assume that farmers adjust their expectations. A positive shock has the opposite effect, causing the farmer to apply more inputs in subsequent periods to help produce an even more abundant crop.

Fig. 2.6 illustrates the results of this multistage production simulation. The first thing to notice here is seasonality, in particular, the big jump in labor demand in the final, harvest period. The other main lesson from this figure is that small shocks in the preharvest period have magnified impacts on the demand for labor at harvest time, because the effects compound one another. You can see that the variance in labor demand increases from one period to the next to the next in the preharvest period. Then there is a big increase in volatility in the final, harvest period, even though the harvest is risk-free in our model, and good and bad weather events *could* potentially counteract one another in the previous five periods.

WHAT DO THE DATA SHOW?

Each quarter the United States Bureau of Labor Statistics carries out a Quarterly Census of Employment and Wages (http://www.bls.gov/cew/data.htm), in which it collects monthly data on labor demand from US employers, including farmers. We can use these employment data to see whether the seasonality and volatility of farm labor demand in the real world looks like what our simulation gave us.

Ideally, we would like to test our model for specific crops or even specific farms, but we only have aggregate data, and only for large categories of products (here, fruits and nuts). Low seasonality for some farms, crops, or regions might cancel out high seasonality for others, and the same might happen with regard to negative and positive deviations in employment. This would make it hard to test our hypotheses that the farm labor demand is seasonal (like in a multistage production function) and that small deviations early in the season lead to larger ones later on. But let us see what the aggregate data tell us. If we find patterns in the aggregate, the pattern is bound to be more pronounced at the regional or farm level.

How might we test the hypothesis that small deviations in labor demand early in the season lead to larger ones later on? One way is to compare year-over-year labor demands, month by month, in a relatively normal year, then in a year in which something caused the labor demand to deviate from its norm. The 2004 and 2005 crop years in California fit the bill nicely, so we will use them.

Fig. 2.7 shows the year-to-year variation in labor demand for California fruit and nut production in each of the 9 main cropping months (April through November). We calculated this variation by taking fruit and nut employment in a given month (say, April 2004) and dividing it by the average April fruit and nut employment over the previous 5 years. If the result is >1.0, it means above-average April employment. Below 1.0 means the opposite.

If our theory is correct, we can expect to see that months in which the year-over-year labor demand is stable (1.0 on the vertical axis) tend to be followed by other stable months, but once employment begins to deviate in one direction or the other the deviations tend to spill over into the next months and magnify.

You can see that things were fairly stable in 2004: Total employment on fruit and nut farms was only slightly higher each month than in the same month during the previous 5 years. These small positive deviations turn into bigger positive deviations in labor demand in the final 2 months of the crop year, though.

The 2005 crop year is a different story. The year-over-year deviations start out positive, then something happens that causes employment to go south. Once employment turns negative, it drops even more. The negative deviations are biggest in the last 2 months—up to 20% below the previous year and nearly 60% below where the positive deviations were earlier in the year. You can see that there are big fluctuations in employment year-over-year. Labor demand drops by as much as 17%, and it jumps by as much as 36%, from 1 year to the next.

Aggregate numbers mask the variability of labor demand on individual farms or in individual farm regions. The problem of seasonality is compounded by the uncertainty of farm labor demand, particularly at harvest time but also at

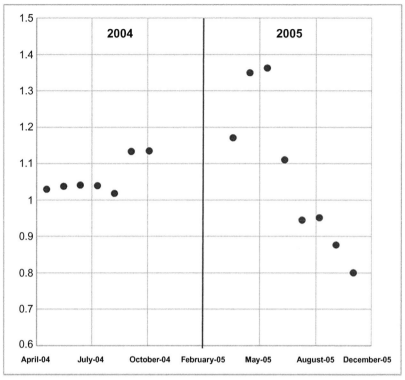

FIG. 2.7 Variations from 5-year trend in California's fruit and nut employment. *(Data from California Employment Development Department; http://www.labormarketinfo.edd.ca.gov/.)*

previous stages of the production process. At each stage, the outcome of input demand decisions is uncertain, or stochastic. As a result, labor demands at subsequent stages of the production process are also stochastic.

WHY SEASONALITY MATTERS SO MUCH

The seasonality of farm labor demand creates challenges for farmers, workers, rural communities, and policymakers.

Farmers must find ways to secure access to large numbers of workers for short periods of time. A typical fruit farm, from California to France, employs a small number of year-round workers to prune, irrigate, and tend to other tasks throughout the year, *plus* a large number of short-term harvest workers. Harvest timing windows tend to be narrow—fruits have to be picked at just the right time in order to be marketable. If farmers cannot find a sufficient number of workers when they need them, harvest losses may result.

If you are a farmer, how do you find the workers you need at harvest time? You could sit tight and hope the workers will come. Workers often return to the same farm to harvest crops year-to-year. Some workers feel an allegiance to particular farms, and some farmers invest effort to strengthen this allegiance. Alternatively, a farmer might rely on a farm labor contractor (FLC) to meet her seasonal labor needs. Then part of the burden of matching labor supply and demand shifts from farmer to FLC.

But what if the crop ripens early? Or late? Many workers migrate from farm to farm, moving northward from one crop to another as the season progresses. FLCs line up contracts among farms, moving crews over space as crops ripen. If the harvest comes late to the south, will migrant workers arrive in time to harvest crops farther to the north? Will the FLC be able to come through for the next farmer? This is a subject of Chapters 3 and 4.

APPENDIX A

Derivation of Farm Labor Demand

Expressing Production as a Function of Inputs

We begin by giving a mathematical form to the lettuce production function. The production function, $Q = F(L, K)$, must follow a few basic rules, which you no doubt learned about in your microeconomics courses. It has to be increasing in variable inputs. In this simple production function, the only variable input is L, so $dQ/dL = MPL > 0$.

Many functions have this characteristic. The simplest is a linear function. However, a linear production function would not satisfy the second rule of production functions: that production exhibits diminishing marginal returns to variable inputs. That is, the rate at which the MPL changes, which is the second derivative, is negative ($d^2Q/dL^2 < 0$).

By far the most widely used production function in economics is the Cobb-Douglas production function: $Q = AL^{\alpha}\overline{K}^{1-\alpha}$, where $0 < \alpha < 1$, A is a positive shift parameter, and the bar over K reminds us that capital is fixed. It satisfies both of the properties we are looking for. Its first derivative is:

$$dQ/dL = \alpha AL^{\alpha-1}\overline{K}^{1-\alpha}$$

It is positive given that α is positive for any positive level of labor demand and output. The second derivative is:

$$d^2Q/dL^2 = (\alpha - 1)\alpha AL^{\alpha-2}\overline{K}^{1-\alpha}$$

It is negative because the coefficient $(\alpha - 1)$ is negative. If we draw a picture of this production function, it will look a lot like the general production function in Fig. 2.1.

Profit-Maximization With a Cobb-Douglas Production Function

We can derive the optimal labor demand using the Cobb-Douglas (C-D) specification. In the textbook C-D production function, profit is given by:

$$\pi = pQ - wL = pAL^\alpha \overline{K}^{1-\alpha} - wL$$

Note that this production function assumes that the total revenue (TR, the value of goods produced) is attributable only to labor and capital. Later on we will have to make an adjustment when we estimate actual crop labor demand, in order to account for purchased intermediate inputs that are also part of a crop's value.

The firm demands variable inputs (here, labor) to maximize.

$$pQ - wL = pAL^\alpha \overline{K}^{1-\alpha} - wL$$

The first-order condition for profit maximization is

$$d\pi/dL = pdQ/dL - w = p\alpha AL^{\alpha-1}\overline{K}^{1-\alpha} - w = 0$$

We can simplify this by substituting in the original production function. We obtain

$$d\pi/dL = p\alpha Q/L - w = 0$$

Solving this for L gives the profit-maximizing labor demand

$$L = p\alpha Q/w$$

You can see that the labor demand increases with the output price and decreases with the wage. It also increases with production.

This labor-demand equation is nice and simple, but bear in mind that Q is not really given; it depends on L. A change in price, wage, or other shock (like weather, which we will explore further) will change Q. If p goes up, L will increase for two reasons: the direct effect of p on L in this equation, and the indirect effect through Q (production increases when the output price rises).

Our simple labor demand function is a useful approximation when we want to explore how exogenous shocks affect labor demand. It also gives us a way to calculate α from real-world data. Solving for α we obtain.

$$\alpha = wL/pQ$$

If we know a farm's total wages (wL) and total revenue (TR $= pQ$), we can divide one by the other to calculate α. Our production function takes L and K and turns them into output.

Data to Construct a Farm Labor Demand Model

Production functions and farm labor demand equations can be estimated with econometric methods, using micro-data from many different farms. But it is possible to construct a farm labor demand model more simply, using averages or data for a representative farm.

The University of California, Davis maintains cost studies for a large number of agricultural crops. These studies are publically available at: http://coststudies.ucdavis.edu. Each study provides cost data on labor and other inputs by production phase. The 2010 cost table for wrapped iceberg lettuce in the California Central Coast counties of Monterey, San Benito, and Santa Cruz appears in Table 2.A1. The table shows a total cost per acre of $9582. It includes labor costs, which we calculate as the sum of hand labor and custom thinning and hand hoeing, for a total labor cost of $2870. The majority of these costs—$2520—are at harvest.

There are also intermediate input costs, which include seed, irrigation, fertilizer, herbicide, and packing cartons. Intermediate inputs total $5512.[2] The single largest component is the cost of packing cartons at harvest time ($2160).

Intermediate inputs are not in our production function. In real life, farms and other firms buy intermediate inputs (seed, fertilizer, etc.) and add value to these inputs by combining them with labor, capital, and other factors. The production function is really a value-added production function. It describes the increase in value above and beyond the cost of intermediate inputs. This means we will have to net out the cost of intermediate inputs from the total value of crop output. It is an example of how things become a bit more complicated when we take theory out into the real world![3]

The share of total value added (*TVA*) in TR is the value-added share (*VASH*). It is the part of the total value of production that is created by the arguments in the production function (*L* and *K*). We can calculate it fairly easily as one minus the share of intermediate inputs (*INTER*) in TR:

$$VASH = TVA/pQ = 1 - INTER/pQ$$

The *VASH* is the share of the output price that is "created" by the labor and capital factors. Multiplying *VASH* by *p* gives a modified price called "the value-added price." The value-added price is the output price with the value of the intermediate inputs netted out of it. This is the price we have to use when doing applied work.

---

2. This can be calculated as total growing costs ($2902) + total harvest costs ($5480) = $8382, minus the total labor costs ($2870).

3. We could have included intermediate inputs in this production function, but we chose not to for a couple of reasons. First, adding additional variable inputs (besides labor) would complicate the model, and we wanted to keep it simple. Second, a Cobb-Douglas production function assumes that the inputs included in the function are substitutable. It makes sense that labor and capital factors can substitute for each other, but it is less clear that labor and, say, seeds or fertilizer are substitutes in growing lettuce. The intermediate inputs in this case study are like ingredients in a recipe, which the cook's labor and capital turn into a high-value meal. The team of experts doing the study determined the list of intermediate inputs and quantities per acre.

TABLE 2.A1 Sample Production Costs and Returns for Wrapped Iceberg Lettuce

UC Cooperative Extension

Sample Production Costs for Wrapped Iceberg Lettuce

Sprinkler Irrigated—40-inch Beds

Central Coast—Monterey, Santa Cruz and San Benito Counties—2010

Operation	No. Times/ Crop/Acre	Cost/Time/ Acre[a]	Material or Custom		Hand Labor[b,c]		Total Cost[c]
			Type	Cost/ Acre	Hours/ Acre	Cost/ Acre	($/ Acre)
Land Preparation and Planting							
Disk	8	20					160
Subsoil (to 32″)	2	45					90
Chisel (to 20″)	2	30					60
Finish chisel	2	24					48
Land plane	1	40					40
Laser plane (1 per 4 veg crops)	0.25	80					20
Lilliston	4	17					68
Harrow and shape beds	2	17					34
Plant	1	25	Seed	150			175
Cultivate and break bottoms	4	14					56
Subtotal							751

Irrigation—Sprinkler							
Preirrigate	2	15	2 ac in	22	1.3	15	67
Irrigate—germination	1	15	3 ac in	33	2.3	25	73
Irrigate—season	5	15	12 ac in	132	11.4	130	337
Subtotal							477
Custom/Contract							
Thin			Custom	100			100
Hand Hoe			Custom	80			80
Subtotal							180
Fertility Management							
Cover crop (1 per 4 veg crops)	0.25	15	Seed	45			15
Soil amendment	1	17	Multiple	75			92
List and fertilize beds	1	15	Preplant	129			144
Side dress beds[d]	2	17	Multiple	264			298
Subtotal							549
Pest Management							
Herbicide (applied at planting)			Herbicide	50			50
Insects and diseases	Multiple		Multiple				525
Pest control advisor							25

Continued

TABLE 2.A1 Sample Production Costs and Returns for Wrapped Iceberg Lettuce—cont'd

UC Cooperative Extension

Sample Production Costs for Wrapped Iceberg Lettuce

Sprinkler Irrigated—40-inch Beds

Central Coast—Monterey, Santa Cruz and San Benito Counties—2010

Operation	No. Times/Crop/Acre	Cost/Time/Acre	Material or Custom		Hand Labor		Total Cost
			Type	Cost/Acre	Hours/Acre	Cost/Acre	($/Acre)
Subtotal							600
Misc Business Costs							
Food Safety/Sanitation							45
Regulatory programs							50
Overhead[e]							250
Subtotal							345
Total growing costs							2902
Land rent and taxes							1200
Total growing costs/Land rent and taxes							4102
Harvest[f]			Cartons/sales				
Harvest, pack, sell @ $5.85/carton				2160		2520	4680

Cool @ $1.00/carton	800
Total harvest costs	5480
Total costs (growing, land rent and taxes, harvest)	9582
Total costs/carton[f]	11.98

UC Cooperative Extension

Sample Net Returns Above Growing, Land Rent & Taxes, Harvest Costs for Wrapped Iceberg Lettuce

Sprinkler Irrigated—40-inch Beds

Central Coast—Monterey, Santa Cruz and San Benito Counties—2010

Yield (Cartons/Acre)	Price/Carton ($)					Breakeven Price
	9	10	11	12	13	
800	−2382	−1582	−782	18	818	11.98
875	−2221	−1346	−471	404	1279	11.54
950	−2060	−1110	−160	790	1740	11.17
1025	−1898	−873	152	1177	2202	10.85
1100	−1737	−637	463	1563	2663	10.58

[a] Includes tractor and driver, fuel, equipment repairs. Tractor labor is $13.40 per hour including 34% benefits/overhead.
[b] Hand labor is $11.40 per hour including 34% benefits/overhead.
[c] Costs per acre can vary considerably depending on many variables including production location and weather conditions, land rent and taxes, soil type, water costs, material inputs, and energy costs.
[d] May include gypsum, compost, water-run fertilizer, and/or soil anticrustant.
[e] Includes interest on operating capital.
[f] Assuming a yield of 800 cartons per acre. Each carton is 24-count and 42-pounds.
Data from University of California Extension, http://coststudies.ucdavis.edu.

The Cobb-Douglas exponent α is the share of labor in value added. Thus, we need to modify the denominator in our formula for α by replacing p with the value-added price, $VASH*p$:

$$\alpha = wL/(VASH \cdot pQ) = wL/TVA$$

Crop total revenue is the product of price p and yield Q. The last panel of the appendix table gives a sense of the range of yields (cartons/acre) and farm gate prices per carton (\$). As we can see in our formula, different prices and yields will result in different estimates of the parameter α. We can take the midpoint of the yield (950 cartons) and price (\$11), for a TR of $pQ = \$11*950 = \$10,450/\text{acre}$.

We obtain total value added by subtracting the cost of purchased intermediate inputs from TR: $TVA = pQ - INT = \$10,450 - \$5512 = \$4938$. The $VASH$, then, is

$$VASH = TVA/pQ = \$4938/\$10,450 = 0.47$$

The Cobb-Douglas labor exponent $\alpha = \$2870/\$4938 = 0.58$. This is also the share of labor in value added. Lettuce clearly is a labor-intensive crop. Of every dollar of value-added generated in Central Coast lettuce production, 58 cents is attributable to labor. The rest is the return on farmers' investment in land (or land rent) and capital (machinery). We cannot disaggregate these without additional information on land and capital costs, which are not available in the report.

Plugging α and $VASH$ into the labor demand equation yields:

$$L = VASH \cdot \alpha \cdot pQ/w = (0.47)(0.58) \cdot pQ/w = 0.27pQ/w$$

This is our fully parameterized lettuce labor demand equation.

With this equation, we can explore the impacts of price and yield shocks on labor demand. The US Department of Agriculture (NASS) estimated an average wage for California field workers equal to \$10.56/h in 2016. The labor demand corresponding to the midpoint lettuce price and yield, then, is given by

$$L = 0.27pQ/w = (0.27)(\$11)(950)/\$10.56 = 269.76 \text{h/acre}$$

The labor demands in Table 2.1 were obtained by varying p and Q over a range of values given in the UC Cooperative Extension report.

APPENDIX B

A Multistage Crop Production Model

The first step in constructing a multistage crop production function is to distinguish preharvest from harvest activities. Let us start with the harvest.

The Harvest Labor Demand

By harvest time, the state of nature has been revealed and the stock of harvestable fruit is on the vine (or in the tree or field). Let us call that stock of harvestable fruit Q_P. (The subscript P means that it was the result of preharvest activities, which we will get to a little later.) Since it is too late to change this stock, it is fixed, just like capital. Harvesting it is basically a resource extraction problem. You do not have to extract (harvest) all of the fruit—only what is worth picking, given the market price and the harvesting cost. The farmer combines harvest labor (L_H) and capital (K_H) to turn the harvestable stock of fruit into a harvested output. Let us call the harvested output Q_H. The production function that describes how applying inputs of labor and capital to the harvestable stock creates a *harvested* output looks like this:

$$Q_H = F_H\left(L_H; \overline{Q}_P, \overline{K}_H\right)$$

By the time we get to the harvest, the harvestable stock, like capital, is given: we cannot change it without going back in time and making different production decisions, or without a different state of nature happening. However, we do not know what Q_P is until we solve the first-stage problem, which we will do shortly.

Like with any production function, we will assume that increasing labor increases output:

$$\frac{\partial Q_H\left(L_H; \overline{Q}_P, \overline{K}_H\right)}{\partial L_H} > 0$$

but at a decreasing rate:

$$\frac{\partial^2 Q_H\left(L_H; Q_P, \overline{K}_H\right)}{\partial L_H^{2}} < 0$$

Eventually, as you apply more harvest labor, diminishing marginal returns set in. The harvest is also increasing in the amount of harvestable fruit on the tree, and this marginal effect might be decreasing, too.

The profit maximization problem at the harvest stage is fairly straightforward for anyone who has taken a microeconomics course. Labor is the variable input, and the farmer hires labor up to the point where an additional unit of labor raises the value of the harvest by an amount equal to the wage. If it costs more to harvest the next bit of fruit on the vine than what the fruit is worth, you will leave it. If it is cheaper to pick it than it is worth, you will hire additional harvest labor.

Here is how this plays out in commercial agricultural economies like that of California: A farmer hires a crew to do a first picking of, say, a melon field. Only the ripe melons are picked. Once more fruit ripens, he might pay for a second picking. But it might not be economically optimal to do a third picking, given the price of melons and harvesting cost. The farmer could pay workers a set piece rate per box of melons they pick, but workers will not want to work at that piece rate if the melons are few and far between. They will not be able to "pick the going wage."

In short, optimization at harvest implies setting the *MVPL* (the output price times the marginal product of harvest labor) equal to the wage:

$$MVPL_H = p_H \underbrace{\frac{\partial Q_H\left(L_H; \overline{Q}_P, \overline{K}_H\right)}{\partial L_H}}_{MPL_H} = w$$

This condition comes from maximizing harvest profit:

$$\pi_H = p_H F_H\left(L_H; \overline{Q}_P, \overline{K}_H\right) - wL_H$$

We take the derivative of the harvest profit with respect to labor and set it to zero, and we get the first-order condition for profit maximization above, *MVPL* equal to wage.

The optimal harvest labor demand, $L_H{}^*$, is found by setting the wage equal to the *MVPL*, which depends on the size of the harvestable crop.

There is something important to keep in mind in this harvest profit maximization problem: we took the harvestable crop as given. In other words, this is a conditional profit maximization problem—conditional upon what is out in the field to harvest. In real life, the stock of fruit on the vine was produced by combining capital, labor, and other inputs during the *preharvest* period, and the outcome was affected by the weather and other random shocks (pests, pathogens, etc.). Now let us turn to the preharvest optimization problem.

The Preharvest Labor Demand

In the preharvest period, the farmer combines preharvest labor (L_P), capital (K_P), and other inputs (which we shall not worry about just now, in order to keep things simple) in order to produce a harvestable stock, Q_P. The technology to do this is described by a preharvest production function:

$$Q_P = Q_P\left(L_P; \overline{K}_P, \varepsilon\right)$$

We include the Greek letter ε (*epsilon*) at this stage to represent the weather and other random shocks that, given a particular application of labor, capital, and other inputs, can result in a larger or smaller harvestable crop. As before, think of ε as representing good shocks, so that the effect of ε on output is positive.

In microeconomic theory of the firm, profit maximization *always* implies hiring inputs up to the point where their *MVPs* equal their prices. That is what we just did to find harvest labor. Farmers do the same thing in the preharvest period. That is, we should end up with this preharvest labor demand:

$$MVPL_P = p_P \underbrace{\frac{\partial Q_P\left(L_P; \overline{K}_P, \varepsilon\right)}{\partial L_P}}_{MPL_p} = w$$

There is a catch, though: we do not have a market price for the harvestable crop—only for the harvested output. What is p_p, then? What are grapes on the vine worth before they are picked? To solve this problem, we need to figure out how farmers value the harvestable crop.

Ex ante—that is, before nature reveals itself, when the fields are planted and being tended to—the size of the harvestable crop is not yet known. There is not a bar over Q_P. The farmer has to make decisions that, once the weather conditions are realized, will produce (hopefully) a harvestable crop, which in turn will be an input into the harvest activity. She is considering two periods, not one, and the outcome is uncertain. How can we model that?

Actually, we already have everything we need. We take the harvest production function, remove the bar from Q_P, and substitute in the preharvest production function. What we get is the production function viewed from the preharvest period, before we know what Q_P is:

$$Q_H = Q_H\left(L_H, Q_P\left(L_P; \overline{K}_P, \varepsilon\right); \overline{K}_H\right)$$

The objective in the preharvest period is to maximize the total (after-harvest) profit. Viewed from the preharvest period, the profit function is.

$$\pi_P = p_H Q_H\left(L_H, Q_P\left(L_P; \overline{K}_P, \varepsilon\right); \overline{K}_H\right) - w(L_P + L_H)$$

We can take the (partial) derivative of this with respect to L_P, set it equal to zero, and we have the first-order condition for profit maximization in the preharvest period:

$$MVPL_P = p_H \frac{\partial Q_H\left(L_H, Q_P; \overline{K}_H\right)}{\partial Q_P\left(L_P; \overline{K}_P, \varepsilon\right)} \underbrace{\frac{\partial Q_P\left(L_P; \overline{K}_P, \varepsilon\right)}{\partial L_P}}_{MPL_P} = w$$

Now we know what the price of fruit on the vine is. It is the term multiplying the marginal product of preharvest labor:

$$p_P = p_H \frac{\partial Q_H\left(L_H, Q_P; \overline{K}_H\right)}{\partial Q_P\left(L_P; \overline{K}_P, \varepsilon\right)}$$

How much would someone be willing to pay for an additional, say, pound of fruit on the vine? The answer is: the final output price per pound, p_H, times the effect on the harvest. The farm gate price of picked strawberries in California was \$1.11/pound between 2007 and 2012. That is p_H. Suppose an additional pound of strawberries in the field results in an additional 0.9 pounds of harvested strawberries—that is,

$$\frac{\partial Q_H\left(L_H, Q_P; \overline{K}_H\right)}{\partial Q_P\left(L_P; \overline{K}_P\right)} = 0.9$$

Then an additional pound of strawberries in the field is worth about \$1.00 (i.e., \$1.11*0.9 = 99.9$\overline{9}$ cents). If an additional pound in the field always translated into an additional 0.90 pounds picked, the preharvest price would be simple to calculate: it would be a constant. However, adding another pound of strawberries to a field probably affects the final harvest differently if the field is already full of strawberries (Q_P is large) than if there are hardly any strawberries in the field (Q_P is small), so it probably is not a constant.

In fact, if we look at the price of fruit on the vine, we can see that it is uncertain. It has ε in it. If the weather is bad, Q_P will be small (given L_P and K_P). This can change the price of fruit on the vine, as we can see in the p_P equation. So can unusually good weather, which results in an unusually large Q_P.

This means that the farmer really has to maximize profit conditional upon her expectations about the weather and other shocks, ε. She might assume that this year will be an average year in terms of rainfall, temperature, and so on. If so, then we could replace ε with its mean, $\overline{\varepsilon}$, wherever it appears in our model. The farmer would then make the preharvest decisions, including about labor demand, so as to maximize final profits in an average-weather year.

REFERENCE

Taylor, J.E., 2010. Agricultural Labor and Migration Policy. Ann. Rev. Resour. Econ. 2, 369–393. http://www.annualreviews.org/doi/pdf/10.1146/annurev-resource-040709-135048.

Chapter 3

The Farm Labor Supply: Who Does Farm Work and Who Does Not?

Apparently, even the invisible hand doesn't want to pick beans.

Stephen Colbert

Farm work is selective. The characteristics of those who do farm work are differ-ent than those who do not. As countries develop, the characteristics of their popu-lations change in ways that take people away from the farm. This chapter deals with how economists think about and model the farm labor supply and how it changes over time. It looks at the forces that lead to a shift of labor off the farm, the challenges this creates for agriculture around the world, and the measures countries and their farmers take to address them.

HOW MUCH TO WORK?

A basic tenet of microeconomics is that people optimize. Firms maximize profits, subject to technology (the production function) and input and output prices. Consumers maximize utility, subject to their budget constraint and the prices of the goods that bring them satisfaction or utility. In Chapter 2 we learned how to derive a labor demand function from farmers' profit maxi-mization. Households supply the labor to satisfy this farm labor demand.

We begin this chapter by showing how to derive a labor supply function from consumers' utility or welfare maximization. This requires thinking about the utility function and budget constraint in new ways. Intuitively, households demand leisure and other goods to maximize their utility. Every hour of leisure they demand represents an hour without work, which would have earned the household an amount equal to the wage, w. Thus, the wage is the *opportunity cost*, or price, of leisure.

By working, households gain income, which they can spend on other goods that bring them utility. However, there is a cost to working more. The oppor-tunity cost of working is the utility loss from not spending time in leisure

The Farm Labor Problem. https://doi.org/10.1016/B978-0-12-816409-9.00003-3

activities, like playing sports, spending time with loved ones, or watching a great TV show.

A household only has a limited amount of time to allocate between work and leisure. Let us call this limited amount of time the household's *time endowment*. In practice, the time endowment can be a tricky thing to define. Of course, it cannot be greater than 24 hours per household member per day. People have to sleep. It should not include children or elderly household members, who are too young or old to work, or people with disabilities that prevent them from working. For them, the tradeoff between work and leisure is largely irrelevant.

What about all that time we spend doing unpaid work—washing dishes, cleaning house, cooking, changing diapers, caring for elderly or infirm family members, and so on and so on? In most countries, that unpaid work falls disproportionately on the shoulders of women, so there is an important gender dimension to unpaid work. If you do a lot of unpaid work, it probably does not make you feel any better knowing that the value of this work is not counted as part of national income. However, if you hired someone to do that housework, it would be!

Our goal here is to do as Albert Einstein suggests: "Make things as simple as possible, but not simpler." So let us keep it simple by assuming that the household's time endowment, T, is something like 24 (hours per day) minus 8 (hours of sleep), which can be allocated to hours worked (H) or hours of leisure (X_l). In equation form:

$$X_l + H = T$$

The household gets utility from leisure and its consumption of other stuff, X_o, which it buys at a price equal to p_o. You can think of a unit of X_o as being like a bundle of different goods, and p_o as the price of that bundle.

The household chooses X_l and X_o to maximize utility $U = U(X_l, X_o)$, subject to a cash income constraint. Its expenditure on X_o cannot exceed the cash it earns by working:

$$p_o X_o \leq wH$$

The household's decisions are also subject to the time constraint. We will assume that the household does not leave any cash or time unspent, so the cash and time constraints hold as equalities. (There is no saving in this model. We could bring savings into the analysis by making this a two-period model, in which saving today increases consumption tomorrow. But let us keep things simple.)

We need to combine the time constraint with the cash income constraint. That way we will have a consumer model that looks more like the one you probably learned about in your first microeconomics course, with a single constraint. We can do this by solving the time constraint for H to obtain.

$$H = T - X_l$$

Then we substitute this for H in the cash income constraint:

$$p_oX_o = w(T - X_l)$$

You can see that the cash the household has to spend on other goods equals the wage times the time spent working, which is the time endowment minus leisure. Rearranging, we can move both of the goods in the utility function over to the left-hand side of the income constraint:

$$wX_l + p_oX_o = wT = Y_f$$

We call this the *full* or *potential income constraint*. Instead of cash income, on the right-hand side we have wT, which is the most cash income the household could possibly have if it spent *all* of its time endowment working. This full income, Y_f, gets allocated between leisure, valued at its opportunity cost, which is the wage, and other stuff, which the household buys at a market price equal to p_o. The full income constraint is the relevant budget constraint when the household can turn its time into income (by working) or leisure (by not working). In real life, the household may have access to unearned income, for example, dividends from stock holdings, interest from savings accounts, cash assistance, etc. If so, these have to be added to wT in order to get full income. We want to keep things simple, though, so let us set these additional complications aside.

Fig. 3.1 shows the full income constraint as a line with a slope of $-w/p_o$ and an intercept on the vertical (X_o) axis equal to Y_f/p_o, which is the most other stuff the household could buy if it turned all its leisure into work time (i.e., if $X_l = 0$, so $H = T$). To get the intercept and slope, solve the full-income constraint for X_o as a function of X_l. The full income constraint hits the horizontal (X_l) axis where

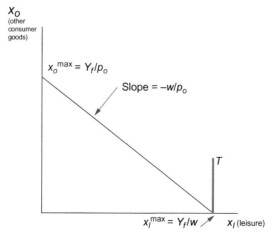

FIG. 3.1 The full income constraint.

the household does not work at all; all of its time is used for leisure. In other words, $X_l^{max} = Y_f/w = wT/w = T$.

Appendix A shows how to mathematically derive the household's optimal labor supply from this full-income-constrained utility maximization problem. At the optimum, the household's marginal rate of substitution between leisure and other goods, $MRS_{l,o}$, just equals the ratio of the wage to the price of other goods, w/p_o. The intuition behind this optimization condition is not so hard to see. The $MRS_{l,o}$ is the rate at which the household trades leisure for other goods subjectively, in terms of its preferences. It is the number of hours of leisure it would give up in order to buy another unit of consumer goods. Working people face this tradeoff all the time. The price ratio is the rate at which *the market* trades off these two goods. It is the number of hours of work a person has to do in order to buy an additional unit of consumer goods. If these two ratios are equal to each other, the household can do no better; otherwise it can. If the household would get more satisfaction from an additional hour of leisure than from what it can buy from an additional hour of work, it will spend the hour in leisure, not work, and vice versa.

We can use a figure to illustrate the optimal allocation of household full income between leisure and other consumer goods. Fig. 3.2 shows this as the

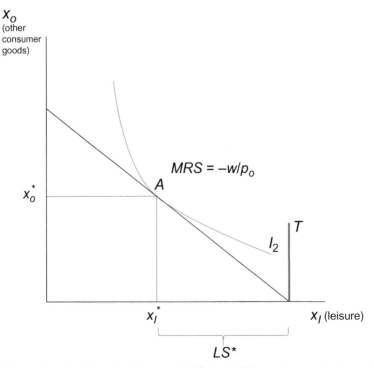

FIG. 3.2   The household's optimal labor supply, LS^*, is the difference between the time endowment (T) and the optimal leisure demand (X_l^*).

point of tangency between the household's indifference curve, I_2, and its full income constraint. You can see that the slope of the indifference curve is -1 times the ratio of the marginal utility of leisure to the marginal utility of other consumer goods, or the $MRS_{l,o}$. It represents the tradeoff between leisure and other goods in terms of the household's satisfaction or utility. The ratio of prices represents the tradeoff between leisure and other goods in the market. The two ratios are equal at point A in the figure. At this point, the household demands X_l^* hours of leisure and an amount of other goods equal to X_o^*.

This figure also shows the household's endowment of time available for work or leisure, T. The household's optimal labor supply, LS^*, is the difference between its time endowment and the optimal leisure demand:

$$LS^* = T - X_l^*$$

This identity comes straight out of the time constraint. Once we know the household's optimal leisure demand, we know its optimal labor supply.

What happens if the wage rises? Figs. 3.3 and 3.4 depict the comparative statics of a wage increase. An increase in the wage has three effects. The first

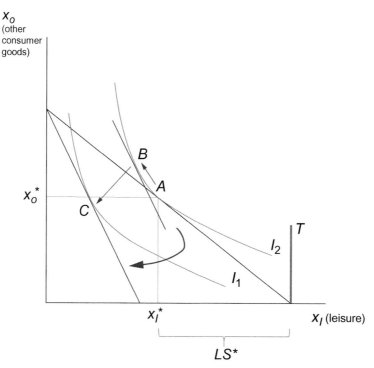

FIG. 3.3 The higher wage pivots the full income constraint. The (compensated) substitution effect moves the household from point A to B, while the real income effect moves it onto a lower indifference curve, from point B to point C.

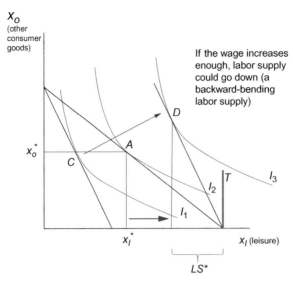

FIG. 3.4 The time endowment effect shifts the full-income constraint outward, from point C to point D, increasing the optimal demand for both leisure and purchased goods.

two, shown in Fig. 3.3, are the same as the effects of any price increase in consumer theory:

1. *The substitution effect.* A higher wage means that leisure becomes more expensive compared to other goods, the price of which (we assume) does not change. At a new wage $w' > w$, the full income constraint becomes steeper, since its new slope is $-\frac{w'}{p_o}$. (Remember: The wage is the opportunity cost of leisure.) Leisure is a consumption good in our model. The household's response is to substitute away from leisure toward other goods, which are less expensive now relative to leisure. If we handed the household cash (i.e., if we compensated the household for the price change) so that it could stay on the same indifference curve, it would shift its consumption from point A to point B of Fig. 3.3. Leisure demand would go down (and the demand for other consumer goods would go up) because of the substitution effect, so labor supply would increase.

2. *The real-income effect.* In real (i.e., price-adjusted) terms, the household becomes poorer when the cost of one of its consumption goods goes up. The negative real-income effect pushes the household down onto a lower indifference curve, from point B to point C in Fig. 3.3. At point C, the household demands less of all normal goods—which in our model include both leisure and other stuff.

The household's optimal labor supply increases because of both the substitution and real-income effects. The substitution and real-income effects reinforce one another, leisure demand falls, so labor supply increases.

But wait, you might wonder, does not the household benefit as a supplier of labor when the wage increases? You are right, and this brings us to Fig. 3.4, which shows:

3. *The labor endowment effect.* The household owns its time endowment, and the value of time increases when the wage goes up. As an owner of time, the household is better off when it can sell its time to the labor market at a higher wage. Recall that the household's full income is given by wT. An increase in w shifts the full-income constraint (with slope $= -w'/p_o$) outward. For example, if the wage doubles, so does full income, and the full-income constraint shifts out from the origin by a factor of two. In Fig. 3.4, the full-income constraint shifts out until it hits the leisure axis at T, like before. This is the point where, after the wage increases, the household allocates all of its time to leisure. The time endowment effect increases the demand for both leisure and other goods, as long as both are normal goods. How much the two demands go up depends on where the new tangency point is. In the example shown in the figure, the household ends up at point D. There, the amount of leisure it demands is more than before the wage increase (X_l^* is higher at point D than at point A), meaning that the household labor supply goes down when the wage rises. The demand for other consumer goods is also higher at point D than at point A.

As this figure illustrates, labor supply does not necessarily increase when the wage rises. It could decrease. Smaller wage increases do not shift the full income constraint as much as in our example, making it more likely that leisure demand is lower after the wage increase. At low wages, households might value the additional income they get from working more than the new price of leisure, such that higher wages increase labor supply. Indifference curves reflecting these preferences would shift upward and outward, and the new point of tangency would be to the left of point A. The higher wage induces a household with such preferences to work more hours, not fewer. Beyond a point, though, if wages continue to increase, households might "cash in" labor time, taking advantage of the fact that they can enjoy more consumer goods as well as more leisure at the new income that higher wages afford them. Then, the labor supply curve can turn back on itself, becoming what labor economists call a "backward-bending" labor supply curve. Starting from a low wage, the quantity of labor supplied increases as wages rise, and eventually, as wages continue to rise, the quantity of labor supplied begins to decrease.

Historically, as countries' incomes and wages go up, time worked per worker tends to get shorter. Comparisons of countries around the world confirm a negative relationship between wages and work time. For example, in 2015 the average Mexican worked 2246 hours total, which translates into more than 43 hours per week, while in Germany, a much higher-wage country, the average laborer worked 1371 hours total, or 26 hours per week. The relationship between leisure and the wage depicted in Fig. 3.3 can become enshrined in law when governments limit hours worked per week (and require overtime pay).

WHERE TO WORK?

Econometric studies have established, beyond a shadow of a doubt, that education and wages are highly correlated with one another. As people become more highly educated, their earnings increase on average. But to get the higher earnings, people usually do not remain in the same location. Wages and the returns to schooling are higher in countries and in production sectors with more capital-intensive technologies that raise labor productivity. They are also higher in industry and services than in agriculture.

The growth of industrial and service employment offers most workers higher wages as well as greater job stability than typically found in agricultural jobs. Education, capital intensity, nonagricultural employment, migration, and wages are related to one another in complex but predictable ways. For example, a study in rural Mexico found that as people's education increases, their labor supply to agricultural jobs decreases, while migration to nonfarm jobs goes up. When nonfarm employment increases, people move off the farm (Charlton and Taylor, 2016).

Once a household or individual decides how much time to work, the next decision is where to work. In real life, a person's set of feasible employment opportunities can affect the decision both of where to work and of how much to work. For example, getting an education might add an urban job with a higher wage to a young villager's employment possibility set.

We can construct a simple model in which a person decides where to work by choosing the job that brings her the highest utility. To simplify things, imagine that the decisions of how much time to work and where to work are taken sequentially. This might not be a bad assumption if the person knows what her employment possibility set looks like and the wages she could earn in different jobs. The problem we wish to solve here, then, is where to work, given that a household or individual already has decided how much to work.

Before we begin, we should make clear what we mean by "where." The decision of where to work might involve location (stay in the village or migrate to the city), sector (work in agriculture, manufacturing, or services), or even which employer to work for. The model we are about to learn, called a "mover-stayer model," is very general; it has been used for all of these purposes. For short, in what follows we will refer to choice of jobs, but we could easily replace "job" or "sector" with "place," "type of employer," etc.

The Mover-Stayer Model

Let U_{ij} represent the utility person i gets by working in job j, which could include earnings, security, or nonpecuniary benefits. Logically, the person will choose agricultural work ($j=A$) over work in the modern (nonagricultural) sector ($j = M$) if $U_{iA} > U_{iM}$, where the subscripts A and M denote the agricultural and modern sector, respectively. In reality, nothing necessitates that agriculture be less modern or technological than other sectors. Commercial agricultural production can be very sophisticated. In journalist Carey McWilliams' words, commercial farms in the western United States are like *Factories in the Fields*. However, in many societies agriculture is the more traditional sector, in which families have worked for generations. We consider only two sectors, "agriculture" and "modern." This simplifies the notation, but we could generalize the model to include multiple sectors.

This decision rule describes a discrete, dichotomous (0-1) choice. We could define a discrete variable, D_i, that equals 1 if person i does agricultural work and 0 if she does modern sector work; that is, $D_i = 1$ if $U_{iA} > U_{iM}$.

What is utility? We need to posit what utility is in order to continue with our economic analysis of farm labor supply.

Is it a wage? Do individuals choose farm work because, for them, it pays a higher wage than nonfarm work? Is it earnings, which depend not only on the wage but also how much time a person works? The hourly wage might be higher in one job than another but the availability of work lower. In that case, earnings are a better indicator of income in each sector than the hourly wage is. Employment security is a major reason some workers give for switching from farm to nonfarm jobs.

Average earnings are universally higher in nonfarm jobs than in farm jobs. But there is no question that for some people farm work is the highest-income option. If you are a fast picker and do not have the schooling necessary to get a factory or office job, your income may be higher picking crops than doing nonfarm work.

If you are reading this book, there is a very good chance that you would not consider picking crops for a living even if it paid the same income as a nonfarm job. Picking crops is hard. Most people associate farm work with drudgery—more so than nonfarm jobs. Consider this story of a woman who picks lettuce in California and Arizona:

> *From 7:30 am to 6:00 pm, Antonieta bags, ties, and boxes lettuce in the fields. Paid on a piece-rate basis, Antonieta can earn up to $11 an hour when working her fastest. The constant repetition of the work is very difficult. She said that after years of this repetitive movement, her arms are constantly in pain. When the pain in her arms became so unbearable, she visited a doctor, who said that he may recommend surgery if the pain continues. The doctor believes the injury is due to the fast pace and repetitive nature of Antonieta's work. Antonieta said that she is always in pain and her arms feel like they are dislocated. Some nights she cannot sleep because of the pain.*
>
> *Despite this excruciating pain, Antonieta continues to work in the lettuce fields because her family could not survive with just her husband's paycheck. She said that as long as her family needs her, she will continue to work in the fields (Farmworker Justice).*

What could be done to make farm work more tolerable for Antonieta and people like her? What if you assigned her a robot to do the repetitive work? She would have to learn how to work with a robot. And it would have to be worth it to the farmer to invest in robots, and to researchers at UC Davis or the John Deere Corporation to invent robots to do the repetitive work. If there are plenty of people like Antonieta willing to pick lettuce at low wages, there will not be much of an incentive for this to happen. But what happens if the farm labor supply diminishes, if workers transition to higher paying, more comfortable jobs in the nonfarm sector? We study two economic models in this chapter that illustrate this transition out of farm work. Later, we see that this transition already occurred in the United States and is now occurring in rural Mexico and in other parts of the world (Chapter 8). We also discuss how the agricultural industry is responding to a diminishing farm labor supply (Chapter 9).

Imagine that you are growing up in a village in Zambia or some other poor country. Like many villages in poor countries, it has no school. Your parents farm the land and raise a few head of cattle, like their parents before them, and theirs before them. Getting an education is hardly an option—it would require leaving the farm and somehow coming up with the money to support yourself and pay for your studies. It is, in essence, outside of your choice set. Literacy is not necessary to grow subsistence crops. There is a certain equilibrium, in which farming is passed on from generation to generation. It is hard to imagine doing anything different.

What could break this equilibrium? Television brings images of affluent life in the city into your village, but that life seems far away. The people who live it are not part of your reference group—or in your mind, your future.

What if the government builds a school in your village? Literacy becomes an option. Some of your neighbors migrate to a construction or factory job in the city, earning a wage that is low by rich-country standards but high—and more importantly, stable—by village standards.

When girls become educated, research shows that they grow up to have fewer children. Families become smaller. With fewer children, parents can afford to invest more time and income into raising each child. Children's human capital—that is, their skills acquired in school, in the family, and in society—increases.

The combination of smaller families, more investment in human capital, and access to nonfarm jobs expands young peoples' opportunity sets and aspirations, stimulating the movement of labor off the farm. This farm labor transition is part of a larger agricultural transformation in developing countries all around the world today.

What factors shape the farm labor supply? To understand the farm labor supply, we need to investigate the conditions that contribute to an individual's decision to work in agriculture and discuss how these conditions evolve as an economy develops. We begin with two influential models that describe the evolution of the farm labor supply at the aggregate level, and then we examine how individual attributes determine whether an individual works in the agricultural sector. Each of these models involves a special case of the mover-stayer model. In the Lewis model, utility and wages are the same. The Todaro model recognizes that the probability of finding work also affects the mover-stayer decision. Both of these models have been used widely to model rural-to-urban migration in countries around the world.

THE LEWIS MODEL: TOO MUCH LABOR ON THE FARM

W. Arthur Lewis, a Nobel laureate, developed one of the most influential models of farm labor supply in his 1954 article, "Economic Development with Unlimited Supplies of Labor." The purpose of Lewis' article is to model the distribution, accumulation, and growth of capital. Since poor countries usually start out with small endowments of capital relative to labor, one can imagine that this article has important implications for economic development. Capital growth also has important implications for labor, since it can make labor more productive. Consequently, the Lewis model provides a good foundation for understanding the evolution of the labor supply from a labor abundant, subsistence agricultural economy to a more capital-intensive, commercial economy with rising wages.

In the Lewis model, at low levels of economic development countries appear to have an almost limitless supply of workers in the agricultural sector, which often is referred to as the subsistence sector. There is such a large number of workers in this sector that the marginal product of labor in agriculture is virtually zero. Lewis argues that the marginal value product of labor can even be negative in some instances, but for the sake of simplicity in our model, we will say that it is slightly more than zero. The marginal value product of labor is equal to the subsistence wage, which is just enough to keep an individual alive. Call the subsistence wage w_s.

Suppose that there are $\overline{L}$ individuals in the economy, and there are two industries: agriculture (the traditional sector) and manufacturing (the modern sector). Everyone works in either the agricultural or manufacturing sector. This economy is represented in Fig. 3.5. The width of the graph equals $\overline{L}$. The curve labeled $MVPL_M$ (read left to right) maps the marginal value product of labor in the manufacturing sector, and $MVPL_A$ (right to left) is the marginal value product of labor in the agricultural sector. The $MVPL_A$ decreases until it reaches zero just to the left of point A. At point C, all workers are in the agricultural sector. Between points C and A, even if we were to remove workers from agriculture and send them to manufacturing, the $MVPL_A$ would not rise, because the farm sector is so saturated with labor.

In this labor-saturated agricultural economy, workers are not paid a wage equal to their marginal value product (since the $MVPL_A$ is zero, this wage would be zero). Instead, workers receive a subsistence wage just high enough to maintain and reproduce themselves. In traditional societies, various institutions exist

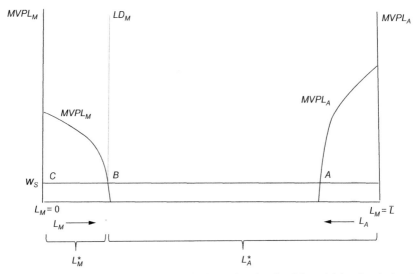

FIG. 3.5 Marginal value product of labor in the modern (reading left to right) and agricultural (right to left) sectors. The marginal value product of labor in agriculture decreases then becomes zero to the left of point A.

to redistribute food and help ensure the survival of a workforce that is larger than can gainfully be employed—that is, a surplus or redundant workforce.

Lewis noted that at the earliest stages of economic development, the economy starts out at point C. Almost the entire population works in agriculture, and almost no one works in manufacturing. As you move along the horizontal axis from left to right, the number of people working in the manufacturing sector increases and the number working in agriculture decreases. Let L_A be the number of workers in the agricultural sector, and let L_M be the number of individuals employed in manufacturing. Anyone not employed in manufacturing is still in agriculture. There is total employment such that $L_A + L_M = \overline{L}$.

This scenario with subsistence earnings in the agricultural sector is not appropriate for describing all economies. However, it can be applied to economies where the population is large relative to capital and natural resources. These circumstances describe many developing economies, especially in Africa and Asia, where birthrates are high and farmers are pushed out into marginal, low-productivity lands. Without sufficient capital to complement the labor supply, the marginal product of labor is small.

Now we introduce modern sector expansion into the model. Owners of capital (referred to as "capitalists") can employ workers by offering them wages equal to their opportunity cost of working in the traditional agricultural sector, which is the subsistence wage. The marginal value product of labor in the modern sector is initially very high and it curves downward as more workers are

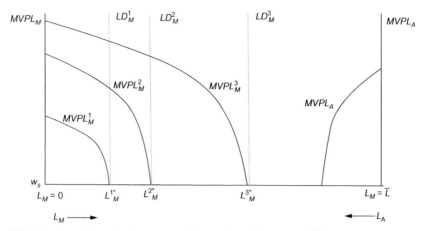

FIG. 3.6   Increasing capital investment shifts out the modern sector *MVPL* curve.

employed. How many workers does the modern sector employ? The capitalists employ workers to the point where the marginal value product of labor is equal to the subsistence wage, point *B* in Fig. 3.5. This is the point that maximizes capitalist profits. Demand for labor in the modern sector in this figure is given by LD_M. The area underneath the $MVPL_M$ curve and above the subsistence wage measures capitalist profits, also called the rents to capital.

What do the capitalists do with these profits? They invest in more capital. As they add new capital to the modern sector, the existing modern sector workforce becomes more productive. Fig. 3.6 shows this as a series of upward and outward shifts in the $MVPL_M$ curve. (The economy will never operate at a wage below the subsistence wage, so we cut this and the following figures off across the bottom at w_s.) The optimal number of workers employed in the modern sector increases, luring more workers off the farm and absorbing some of the labor surplus. With each addition of capital to modern-sector production, the rents to capitalists increase, generating greater incentives to continue investing in capital. The cycle of capital investment draws additional workers from the traditional sector to the modern sector, increases profits and savings, then leads to new rounds of capital investment.

Capital investment is not limited to the modern sector. As the modern sector grows, the demand for food rises. The modern manufacturing sector does not produce its own food, but people have to eat, and rising incomes stimulate food demand as people move off the farm. This creates incentives for farmers to invest in agricultural capital and new agricultural technologies, including higher-yielding seeds. More capital and more productive seeds push the $MVPL_A$ curve upward and outward, as illustrated in Fig. 3.7.

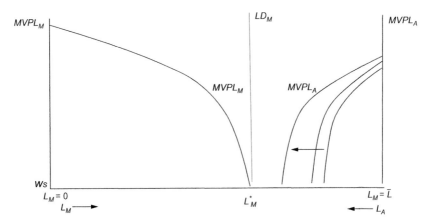

FIG. 3.7 Technology change in agriculture pushes out the *MVPL$_A$* curve.

Where does this process end? Eventually, after many iterations of the rents-savings-investment-hiring cycle, the two MVPL curves intersect and the economy reaches what is called the "Lewis turning point." The Lewis turning point is at L_M* in Fig. 3.8. It corresponds to the level of capital investment at which the *MVPL$_M$* curve intersects the *MVPL$_A$* curve at the subsistence wage. To the right of this point, agricultural workers can produce a value greater than the subsistence wage. Because of this, the reservation wage, or wage required to induce an additional worker to move into the manufacturing sector, rises as the manufacturing sector expands further. Labor no longer appears as an infinitely abundant resource.

The intersection of the two *MVPL* curves determines an equilibrium wage, which is the wage at which the total quantities of labor demanded and supplied in the economy—and in each sector—are equal. It is the wage employers in both

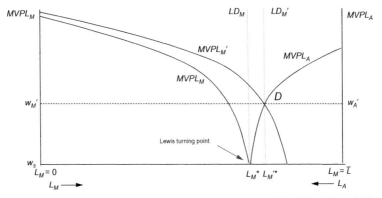

FIG. 3.8 The Lewis turning point occurs where the manufacturing and agricultural marginal product of labor curves intersect above the subsistence wage. If manufacturing expands beyond this point—say, to *MVPL$_M$'*—the equilibrium wage rises above w_s.

sectors have to pay their workers. Employers in both sectors hire labor where its *MVP* equals this equilibrium wage. If either sector wishes to expand its demand for labor further, the equilibrium wage will rise for both sectors. When economies pass the Lewis turning point, they morph from a classical world of labor surplus into a neoclassical world of labor scarcity.

To the right of the Lewis turning point, like at point *D* in Fig. 3.8, the once infinitely elastic or "flat" supply of labor steepens. Wages in both sectors rise and workers begin to capture a larger share of value produced in each sector. The standard of living rises. At point *D*, the equilibrium wage, $w_M' = w_A'$, is well above the subsistence wage.

Profits to capitalists are equal to the area under the $MVPL_M'$ curve and above the horizontal line corresponding to the equilibrium wage. Workers in the manufacturing sector collect total wages equal to $w_M' * L_M'^*$, while workers in the agricultural sector collect total wages equal to $w_A' * L_A'^*$, or equivalently $w_M' * (\overline{L} - L_M'^*)$, since wages are equal in the two sectors. Passing the Lewis turning point is clearly beneficial for workers, and it is an important stage of development as workers become more productive.

The Lewis model makes many generalizations and simplifications about the population and the economy, but it also describes an important transition from farm to nonfarm work and a rise in wages that we observe globally. As economies develop and incomes rise, workers shift out of agriculture, and the farm labor supply progressively becomes more inelastic. The Lewis model helps us understand why and how this occurs. However, a single model is not always sufficient to understand the labor decisions that shape an economy.

The migration of labor underlies the Lewis transition, because gaining employment in the manufacturing sector usually requires people to move out of rural areas. This leads us to new questions: How do individuals decide whether to migrate? What are the costs of migration? What do potential migrants know about the urban sector? Lewis' model assumes that, beyond the Lewis turning point, manufacturing wages are equal to the *MVPL* in the agricultural sector. But in real life, wages typically are much higher in the manufacturing sector, and rural-to-urban migration continues despite high urban unemployment. How can we account for a wage premium in the manufacturing sector? Is it simply what is needed to compensate rural people for costs of moving to the city? And why do millions of people in less-developed countries migrate to cities where unemployment rates are high?

MIGRATION AND UNEMPLOYMENT: THE HARRIS-TODARO MODEL

John Harris and Michael Todaro (Harris and Todaro, 1970) wanted to answer a puzzling question: Why do we see high and accelerating rural-to-urban migration in countries where urban unemployment rates are high? Why would someone migrate to the manufacturing sector if there is unemployment in that sector? Is not a subsistence wage better than no wage at all?

The Harris-Todaro model (often called simply "the Todaro model"[1]) describes migration across two sectors of an economy when there is unemployment in one or both sectors. Migration into destinations with high unemployment at first glance seems counterintuitive. However, if individuals maximize *expected earnings* rather than wages, migration from farms to cities with high wages and high unemployment can actually make sense. Harris and Todaro show that it is consistent with rational behavior.

In Lewis' (1954) model, past the Lewis turning point the manufacturing wage is equal to the marginal value product of labor in the agricultural sector, so individuals are indifferent between working on the farm and working in manufacturing. There is no unemployment because the manufacturing sector employs only as many workers as it needs so that the marginal value product of labor in manufacturing is reduced to the marginal value product of labor in the agricultural sector.

We would expect market-clearing wages in the urban sector to be greater than the marginal value product of labor in the rural sector by the amount it costs individuals to migrate. Individuals will not migrate to the city if wages are equal across locations and they have to pay a bus fare or other costs to get to the city. We might expect nominal or cash wages in the city to be greater than the nominal wage in rural areas if the cost of living is greater in the city. Even if we adjust urban wages for cost-of-living differences (i.e., use real instead of nominal wages), though, urban wages are higher on average than rural wages. Labor markets do not clear—there is not full employment in the economy—if real urban wages, net of migration costs, exceed real rural wages.

Harris and Todaro argued that minimum wage laws, union activity, and other considerations outside of (exogenous to) the interactions of supply and demand set the urban wage at a level above the market-clearing wage. This creates urban unemployment. They designed a model to understand the impacts of high urban wages and unemployment on migration decisions.

The Harris-Todaro model assumes that the marginal value product of labor in the agricultural sector is greater than zero, and it is decreasing with the addition of more agricultural workers. This places the Harris-Todaro model to the right of the Lewis-turning point. The critical assumption in the Harris-Todaro model is that rural-urban migration will continue as long as the expected urban real wage at the margin is greater than the real agricultural wage (plus migration costs).

---

1. The original article, written by Michael P. Todaro in 1969, is "A model of labor migration and urban unemployment in less developed countries" (*The American economic review* 59, no. 1:138–148.) A year later, Todaro teamed up with John R. Harris to present a more formal treatment of the model, in "Migration, unemployment and development: a two-sector analysis" (*The American economic review* 60, no. 1:126–142). Together, these two articles had over 13,000 citations as of 2018, according to Google Scholar.

What is the expected urban wage? It is not the actual urban wage because there is unemployment in this model. Everyone who migrates to the urban center and obtains a job receives the real urban wage. But the probability of landing a job is not 1. The model assumes that the probability of finding an urban job is equal to the urban employment rate or ratio of jobs to the size of the urban labor force. This would be the case if jobs were given out randomly, like in a lottery. (Other researchers proposed refinements of the model in which employment is not random.) With random employment, the expected wage is equal to the urban wage times the share of the urban workforce actually employed. When there is unemployment, this is smaller than the actual urban wage.

What implications can we draw from the Harris-Todaro model for understanding rural-urban migration and the farm labor supply?

First, we can expect workers to migrate to the urban sector even when there is high unemployment in the urban sector, as long as the expected urban wage is greater than the marginal value product of labor in the rural sector. (This implicitly assumes full employment in the rural sector, i.e., that the probability of finding farm work is one—though the model can be modified to include rural unemployment, too.)

Second, the model shows that the equilibrium allocation of labor to the rural and urban sectors is not efficient as long as there is urban unemployment. Whenever the urban employment rate is less than one, at equilibrium the marginal value product of labor in the urban sector is greater than the marginal value product of labor in the agricultural sector. An efficient allocation would imply equality of the marginal value product of labor between the two sectors.

Third, the creation of an additional job in the urban sector raises expected urban wages by increasing the employment rate. To restore equilibrium, more people migrate from the rural to the urban sector. Thus, urban employment begets migration, which swells the urban workforce. Mathematically, if the new migration is greater than the number of new urban jobs created, urban unemployment will increase. A government project that increases urban employment, therefore, could increase urban *un*employment! Harris and Todaro contend that the creation of one additional job in the urban sector induces the migration of more than one individual from the rural sector.

Appendix B presents the Harris-Todaro model in greater mathematical detail. The major contribution of Harris and Todaro is to show how an urban minimum wage greater than the market-clearing wage affects migration. It explains why we often see increasing rates of rural-urban migration even when there is high urban unemployment, and demonstrates how migration can serve as an equilibrating force when wages are held constant at nonmarket levels. We observe these phenomena around the world, especially as countries develop and urban centers begin to expand.

The Harris-Todaro model assumes that the probability of obtaining an urban job is the same for all potential migrants, and it assumes that the opportunity cost of migration is equal for all workers. In reality, we know that individuals differ from one another, whether in their natural capacities to learn certain skills or tasks or in the experiences and skills that they have acquired in life, including at school. These skills and life experiences affect individuals' productivity in each sector, their preferences, and ultimately their expected wages.

Understanding how individual characteristics affect migration and sector choice is a critical element in understanding changes in the farm labor supply. For this, we need to return to the mover-stayer model.

THE MOVER-STAYER MODEL WITH HETEROGENEOUS INDIVIDUALS

The Lewis and Harris-Todaro models illustrate complex changes in the farm labor supply at the aggregate level, but the farm labor supply is a collection of people, each with his or her unique set of skills and preferences. We cannot observe all of the characteristics that determine who a person is and how s/he makes decisions. But we can make some observations that help us understand how individuals generally make decisions given their observable characteristics. A micro-level mover-stayer model is useful to understand trends in farm labor supply, as well as how changes in characteristics of people and economies shape these trends.

To model who migrates to the manufacturing sector and when, we must return to the question that we asked earlier: what is utility? We cannot observe utility directly because it is an abstract term to describe individuals' or households' welfare or preferences across a set of goods or decisions. But we can generally assume that utility increases with expected income.

Harris and Todaro in a Mover-Stayer Model

Assume that each worker i chooses the job in which his or her expected full income, Y_i, is highest. As before, denote the agricultural sector by the subscript A, and denote the urban nonfarm, or manufacturing, sector by the subscript M. Then we can define individual i's discrete choice to work in agriculture as:

$$D_i = 1 \text{ if } Y_{i,A} > Y_{i,M}, \ 0 \text{ otherwise}$$

Under this specification, $D_i = 1$, meaning that individual i works in agriculture, if her expected agricultural income is greater than her expected income in the manufacturing sector. Otherwise $D_i = 0$.

What determines expected income? Harris and Todaro's answer is anything that affects the probability of finding a job and the wage once a person is employed. Human capital affects both the probability of employment and wage, for two reasons. First, human capital raises labor productivity. The microeconomic theory of the firm posits that workers are paid the marginal value product of their labor. When people raise their education, their wages rise because they become more productive. Second, with higher education, individuals gain access to more stable jobs in factories, construction, businesses, offices, etc.

Employers often only consider hiring workers with education at or above some minimum.

The impact of schooling on nonfarm income is potentially much larger than the impact on farm income. Some nonfarm jobs do not require much education. Urban slums are filled with informal jobs that require little if any schooling at all: the ragpickers of India, who wade through garbage dumps; the people who load boxes of produce onto trucks; the unskilled construction workers who carry off debris from demolished buildings. But a little schooling can go a long way toward moving up the urban job ladder—much more than moving up the agricultural job ladder. We can put this in our model by making expected income depend on a person's education. Let $educ_i$ be the years of education of individual i.

Education is not the only variable that affects expected income. Age, experience, and gender likely affect earnings. Household variables like wealth, family size, and the size and quality of landholdings might also affect potential earnings. In the farm sector, individuals often work on their families' farms. If the family does not have access to perfectly functioning labor markets, then the opportunity cost of time, or family wage, will depend on the household's production capacity along with its preferences for leisure and other goods (as we shall see in the agricultural household model in Chapter 5). When labor markets exist but do not function well, potential earnings may be determined by local wages or productivity on the family farm. The individual will only work on a neighbor's farm if the wages his neighbor pays are greater than or equal to his marginal value product of labor on the family farm. The size of the household, the farm size, and the quality of land help determine the individual's marginal value productivity of labor on the farm.

Let X_i be a vector of variables that include individual, farm, and household characteristics likely to affect the individual's expected farm and nonfarm earnings.

Beyond individual and household farm characteristics, the people that an individual knows may positively influence his potential employment and earnings in a given sector or location. Social networks play a significant role in hiring and earnings, especially for individuals who migrate away from their homes to work. Empirical studies find that migrants who have strong networks with friends or family at migrant destinations are less likely to be unemployed and more likely to have high earnings compared with migrants who do not have access to these networks. Literature suggests that networks are location and sector specific (Davis, Stecklov and Winters, 2002; Guilmoto and Sandron, 2001; Mora and Taylor, 2006; Richter and Taylor, 2007), so i's network in the urban sector is different from his network in the agricultural sector.

Even in the agricultural sector, close to home, a network may be important. Having close relationships with one's neighbors may increase employer trust, and the employer may be willing to pay a higher wage because he does not need to monitor his friend as closely as he would a stranger. We can denote the individual's network of contacts in the sector of desired employment by $net_{i,j}$, where j represents the sector of employment.

Thus, we can express expected gross earnings for individual i if he works in the agricultural sector as

$$Y_{i,A} = f(educ_i, X_i, net_{i,A}) + \varepsilon_{i,A}$$

where $f(educ_i, X_i, net_{i,A})$ is a function converting individuals' human capital, migration capital, and other characteristics into expected agricultural income, and $\varepsilon_{i,A}$ is an error with mean zero that captures all of the unobservable factors that affect expected farm earnings.

Expected gross earnings in the nonfarm (manufacturing) sector can be represented as

$$Y_{i,M} = g(educ_i, X_i, net_{i,M}) + \varepsilon_{i,M},$$

where $g(educ_i, X_i, net_{i,M})$ is a function determining expected nonfarm earnings, and $\varepsilon_{i,M}$ is an error with mean zero.

We have listed a number or factors that are likely to determine an individual's total expected income in either sector, but in reality we are not interested in total expected income as much as expected income net of the cost of getting to the job. What if nonfarm jobs are located far from home, or transportation to the city is costly? Wages paid to nonfarm workers in the city may be greater than expected earnings at home, but the cost of migrating to the city could be prohibitive.

We can denote the cost of migration by $C_{i,j}(X_i, net_{i,j})$. Notice that the cost of migration depends on individual and household characteristics as well as on the individual's network to the sector of interest. Certain household characteristics, including location and access to private or public transportation, may make travel more or less affordable. A network of friends who commute to the city together may reduce travel cost.

For the migrant, a network can be critical when considering the cost of migration to a distant location. Moving to a new city without knowing anyone and without having access to information is daunting. Most migrants have a limited supply of money to live off of while they search for work and for a home. Friends or relatives who help a migrant through this process can greatly reduce the costs of migration and increase overall expected earnings. Friends or relatives in the city may provide the individual with a less expensive place to live. They may help the individual find a job, reducing search costs.

Considering costs, expected net earnings from working in agriculture can be expressed as

$$y_{i,A} = f(educ_i, X_i, net_{i,A}) - C_{i,A}(X_i, net_{i,A}) + \varepsilon_{i,A}$$

Likewise, the expected net earnings from working in manufacturing can be expressed as

$$y_{i,M} = g(educ_i, X_i, net_{i,M}) - C_{i,M}(X_i, net_{i,M}) + \varepsilon_{i,M}$$

Now that we have identified expressions for the expected net earnings from working in either sector, we can create an expression for the probability that an individual works in the farm sector, $Pr(D_i)$. We expect an individual to work in the farm sector only if his expected net earnings in the farm sector are greater than his expected net earnings in the nonfarm sector, that is, if

$$Y_{i,A} - C_{i,A}(X_i, net_{i,A}) + \varepsilon_{i,A} > Y_{i,M} - C_{i,M}(X_i, net_{i,M}) + \varepsilon_{i,M}$$

We can find the probability that the individual works in the farm sector according to the probability function

$$Pr(D_i = 1) = Pr[f(educ_i, X_i, net_{i,A}) - C_{i,A}(X_i, net_{i,A}) + \varepsilon_{i,A}$$
$$> g(educ_i, X_i, net_{i,M}) - C_{i,M}(X_i, net_{i,M}) + \varepsilon_{i,M}]$$

How do an individual's characteristics affect $Pr(D_i = 1)$? The economic returns to specific skills are likely to differ across sectors. The skills that raise labor productivity in agriculture often are not the same as the skills that raise productivity in the manufacturing sector. Skills developed in school may be better suited for the nonfarm sector than the agricultural sector. If so, then even if going to school makes a person more productive in both sectors, the impact on a person's earnings will be smaller for farm than nonfarm work: $\frac{\partial f}{\partial educ_i} < \frac{\partial g}{\partial educ_i}$. Higher schooling then will reduce the probability of working in agriculture.

How might we test this hypothesis and measure the impact of education on the probability of working in agriculture? We might compare the work outcomes of individuals with different levels of schooling. But individuals differ in other ways besides schooling, including ways that we cannot observe. For example, children who attend more school might have different innate abilities that also make them more productive in nonfarm work. If so, a person's ability will correlate both with her education and with her sector choice, and it will be difficult to disentangle the causal effects of schooling from ability.

Sometimes researchers observe natural experiments that generate exogenous changes in education among children and their peers, thus allowing them to measure the impacts of education on different outcomes. Some economic studies use construction of schools in villages to predict the effects of education on wages and other outcomes. For example, the Indonesian government made large investments in school construction in rural areas over a short span of time in the 1970s. It built more than 61,000 new primary schools between 1973–74 and 1978–79. In villages where children did not have access to schools, suddenly they did. Esther Duflo (2001) took advantage of this exogenous rise in school supply to measure the impact of education on wages.

This study was like an experiment, in which the treatment group was children who were primary school aged when the new school appeared in their village, and the control group was children in the same village who were too old to take advantage of the new school. Theoretically, children from the same village who turned school age just before the school construction began should be no different from children who were school age just after construction, except for their access to the schoolroom.

Duflo found that children in the treatment group received significantly more years of education than children from the same village who were too old to benefit from the school construction program. On average, children in the treatment group also had higher earnings as adults than children in the control group.

Duflo's study suggests that improvements in school supply raise wages. How else might improvements in schools affect rural economies? Might they affect the farm labor supply? Our intuition suggests that improved education would reduce the probability that individuals work in agriculture. How might we test this hypothesis?

Suppose for the time being that our hypothesis is correct, that increasing an individual's education reduces the probability that the individual works in agriculture. If this hypothesis is true, then once schooling begins to increase, it might unleash a process of what sociologists call *cumulative causation* (Massey et al., 1993).

Here is how it works: A few kids with schooling migrate to the city to work in the nonfarm sector. Now the network of friends working in the nonfarm sector expands for everyone in the village. This reduces the probability of working in agriculture for all of the kids' friends. Raising education for one individual reduces the probability that the individual works in agriculture, and it also reduces the probability that any of his friends work in agriculture. Changes in parameters that affect the expected earnings in the farm and nonfarm sector for just a few individuals can have precipitous effects on the farm labor supply at the aggregate level when networks connect individuals' decisions.

APPLYING THE MOVER-STAYER MODEL TO DATA

The theory we have learned in this chapter leads us to the hypothesis that people move out of farm work as their economic prospects in other activities improve. Schooling plays a key role in opening up employment opportunities outside agriculture. How can we empirically test the hypothesis that more schooling lowers people's probability of doing farm work?

The decision to do farm work or not is a discrete (0–1) choice. We could survey a random sample of people in a country, asking each one first, how many years of schooling s/he has, and second, whether s/he works in agriculture or in a nonfarm job. We could record the answer to the second question by giving a "1" to a person who did farm work and a "0" to a person who did not. We would want to do this on a large random sample of people—several hundred at least, if not thousands. We could then use econometrics to estimate the correlation between an additional year of schooling and the probability of working in agriculture.

To illustrate this method in the simplest way possible, suppose we interviewed only two people. Person A reports having 3 years of education and

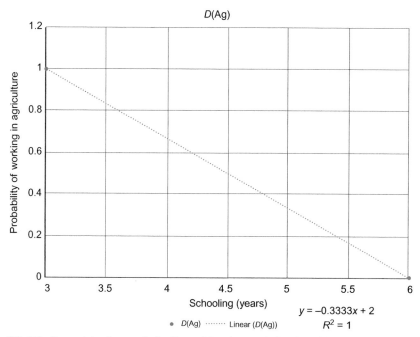

FIG. 3.9   Person A has 3 years of schooling and does farm work ($D=1$), while person B has 6 years of schooling and works outside of agriculture ($D=0$). The slope of the line connecting these two points is −0.33.

working in agriculture ($D=1$). Person B has 6 years of education and works in a nonfarm job ($D=0$). We could graph their responses as a scatter plot of the 0–1 farm labor variable against the years of schooling variable, as in Fig. 3.9.

If we draw a line through these two points, its slope is $1/(3-6)=-1/3$. This is our best estimate of the relationship between schooling and the probability of doing farm work. It is the regression line that we actually would estimate using the ordinary least squares method in econometrics. From this very simple example, each additional year of schooling is associated with an estimated decrease in the probability of doing farm work of 1/3, or just over 33%.

It would not be credible to do this study with only two data points, of course. The addition of even a single person to this sample could radically change the estimate. In real life, some people with more than 6 years of schooling will be found doing farm work, and some people with very little schooling will be observed doing nonfarm work. We are more likely to see a scatterplot like the one in Fig. 3.10, which is from a fictitious sample of 15 individuals (which still is a very small sample). We can no longer fit a line that runs through all of the points, but we have a much more solid basis for estimating the relationship between the two variables.

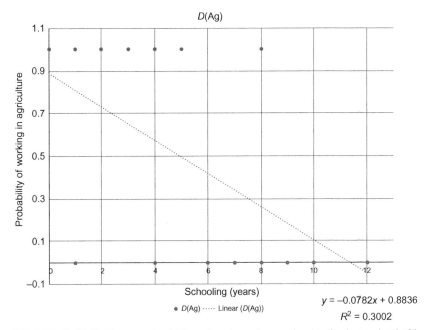

FIG. 3.10 In this fictitious sample of 15 people, a 1-year increase in schooling is associated with an estimated 7.8% decrease in the probability of doing farm work.

We can use the ordinary least squares method from econometrics to fit a line to the observed data points by minimizing the sum of squared vertical distances from each point to the line. This is the best line describing the relationship between schooling and farm work for this sample of individuals. Using these fictitious data, we would end up with the regression line shown in the Fig. 3.10. Its slope is −0.078, which tells us that, on average, a 1-year increase in schooling is associated with an estimated 7.8% decrease in the likelihood of doing farm work.

Fitting a line to these data is not entirely satisfactory for a couple of reasons (besides the fact that the sample is small and fictitious). If you look at the lower-right end of the line you will see that it drops below zero for high levels of schooling; however, it does not make sense for anyone to have a negative probability of doing farm work. You also may notice that the errors, or vertical differences between the data points and the regression line, are not random. They start out being mostly positive (the data points lie mostly above the regression line for low schooling levels) then turn mostly negative (the regression line overpredicts the likelihood of doing farm work for people with higher schooling levels).

In practice, researchers use more sophisticated econometrics and much larger datasets to estimate the relationship between schooling (and other

TABLE 3.1 Estimated Impacts of Age, Gender, Schooling, and Industrial Employment Growth on Rural Mexicans' Probability of Doing Farm Work

Explanatory Variable	Coefficient Estimate[a]
Dependent variable: $D_{(i)} = 1$ if person i did farm work, 0 otherwise	
Age	0.04
Female (1 if female, 0 if male)	−3.27
Schooling (years completed)	−0.16
Industrial employment growth ($\times 100,000$)	−0.20
Observations	65,476
R^2	0.802

[a]All estimated coefficients shown are significantly different from zero at below the 0.01 significance level. This regression controls for a number of other variables; see Charlton and Taylor (2016). Taken from Table 6 in Diane Charlton, Edward Taylor, J., 2016. A declining farm workforce: analysis of panel data from rural mexico. Am. J. Agric. Econ. 98(4), 1158–1180.

variables) with people's probability of doing farm work. Table 3.1 shows the estimated effects of a few key variables on rural Mexicans' probability of doing farm work, controlling for other variables included in the analysis. They are from a study that included 65,476 person-years of data. Controlling for all of the other variables in the study (not shown), the probability of doing farm work increases significantly with age (older people are more likely to do farm work than younger people, at a rate of 0.04% per year of age), schooling (the likelihood of doing farm work falls by 0.16% per year of schooling); and Mexico's industrial employment (an additional 100,000 industrial jobs reduces rural Mexicans' probability of doing farm work by 0.2%). Women are 3.27% less likely to do farm work than otherwise similar men.

Econometric results like the ones in Table 3.1 for Mexico can be found for countries around the world. As countries develop economically, education increases, nonfarm employment grows, and families become smaller. All of these trends associated with economic development decrease the supply of workers available to do farm work.

CONCLUSION

This chapter reviews several models that give us different perspectives to understand the farm labor supply. It is clear from these models that the farm labor supply is not stagnant, and it plays an essential part in an economy's development.

The Lewis model demonstrates how the farm labor supply shrinks with economic development and capital investment. Capital investment in the manufacturing sector draws workers out of agriculture, and capital investment in the agricultural sector can increase worker productivity. Even in populations where farm labor initially is abundant, the industrial and agricultural sectors eventually compete for the same supply of labor, raising worker wages. A great deal of research in development economics focuses on how economies can reach the Lewis turning point, where workers produce value greater than a subsistence wage. This turning point clearly has important implications for increasing worker welfare.

Location is essential in understanding the farm labor supply. Farm labor is found mostly in rural, sparsely populated areas. Leaving agricultural work often requires moving out of rural areas. Changes in the farm labor supply are nearly always linked with rural-to-urban migration. The Harris-Todaro demonstrates how high urban wages draw individuals out of farm work. Even when many workers in the urban sector are unemployed, we can expect to see high rates of rural-urban migration if the minimum urban wage is set high enough. Harris and Todaro show that actual wages do not determine migration; rather, expected wages do. Individuals migrate out of farm jobs and out of the rural sector if their expected urban-sector earnings, which account for the probability of employment, are greater than their earnings in the rural sector.

Finally, we looked at migration and sector choice as an individual decision in a mover-stayer model. Ultimately, it is the individual, often influenced by his or her family, who decides where to work. All individuals are different. Some have more to gain from migration and some have more to lose. As an example, we see that educational opportunities and human capital formation can have large effects on the farm labor supply. Networks with friends and contacts in different sectors and locations are also crucial to the farm labor supply, because they influence a person's expected economic returns from staying or migrating. So, potentially, are the activities of labor recruiters.

In this chapter we explored a baseline model wherein individuals choose their sector of work so as to maximize expected income. But for most individuals, there are other considerations in selecting a sector and location of work. People might seek nonfarm work to gain income for an investment they wish to make back home in the village, or they might use nonfarm work as a form of insurance to protect their household against risks associated with agricultural shocks. Different motivations for seeking nonfarm work can lead to different behavioral responses to environmental or policy changes. We could create models that account for these other considerations. Each model would likely contribute something different toward our understanding of the farm labor supply.

In summary, the farm labor supply is dynamic and complex. It evolves over time as urban economies grow; rural people become more educated; improved transportation, information, and markets increase people's mobility; and people's expectations and preferences change. As people move off the farm, agricultural labor markets evolve, and producers and governments must confront the challenge of producing more food with fewer domestic workers.

APPENDIX A

Mathematical Derivation of Household Labor Supply

Using the variables defined in the text, the demand for leisure and other consumer goods is derived by solving the Lagrangian equation:

$$\max_{X_l, X_o} \mathcal{L} = U(X_l, X_o) - \lambda(wX_l + p_oX_o - wT)$$

In this equation, we can see the utility function as well as the full-income constraint. The Greek letter λ denotes the shadow value on the full-income constraint. It is the marginal utility of income, or utility gain from increasing full income by a small amount. You can also think of λ as an exchange rate that turns full income (measured in, say, dollars) into utility (measured in "utils").

Maximizing the Lagrangian yields the first-order conditions for constrained utility maximization, subject to the full income constraint:

$$\frac{\partial U}{\partial X_l} - \lambda w = 0$$

$$\frac{\partial U}{\partial X_0} - \lambda p_o = 0$$

$$\lambda(wX_l + p_oX_o - wT) = 0$$

These conditions make sense. The household demands leisure and other goods up to the point where their marginal benefit (marginal utility) just equals their marginal cost (price times marginal utility of income, λ). According to the first condition, households demand leisure up to the point where the utility of an additional hour of leisure time ($\partial U/\partial X_l$) just equals the utility loss from not spending that additional hour at work, instead, and earning a wage (λw). The second condition states the same for other goods. The third condition restates the full-income constraint. Unless $\lambda = 0$, the household exhausts all of its full-income demanding leisure and other stuff. We assume that more full income is always better in utility terms, so $\lambda > 0$.

Rearranging, we get the familiar optimizing condition that the marginal rate of substitution between leisure and other goods, $MRS_{l,o}$, equals the ratio of the wage to the price of other goods:

$$-\frac{\dfrac{\partial U}{\partial X_l}}{\dfrac{\partial U}{\partial X_O}} = MRS_{l,o} = -\frac{w}{p_o}$$

APPENDIX B

The Harris-Todaro Model in More Mathematical Detail

For the sake of notation, assume that the urban sector produces manufactured goods and the rural sector produces agricultural goods. Let $\overline{w_M}$ be the minimum wage in the urban (manufacturing) sector, which is set by the government, and assume that $\overline{w_M}$ is greater than the $MVPL$ in the rural sector. Let N_M be the number of workers employed in the manufacturing sector and let N_U be the number of working-age residents in the urban sector (whether they are working or not). Assume that the total endowment of labor in the economy is fixed such that $N_A + N_U = \overline{N}$, where N_A is the workforce employed in the agricultural sector and $\overline{N}$ is the total labor endowment, which can be distributed across both sectors. Denote the expected wage of someone contemplating migration from the rural to the urban sector as w_U^E.

The expected urban wage is the manufacturing wage, set by the government, times the probability of obtaining employment. Then the expected urban wage can be expressed

$$w_U^E = \overline{w_M} \frac{N_M}{N_U}$$

That is, the wage that a prospective rural-urban migrant can expect to receive if he migrates is equal to the fixed minimum wage times the employment rate in the urban sector. As the number of manufacturing jobs increases, the probability of obtaining a job increases and the expected wage increases. As the number of migrants to the urban sector rises, then the probability of obtaining a job decreases and the expected wages decrease.

We are interested in finding how many individuals migrate from the rural to the urban sector. To understand migration, we also need to know the expected wage in the agricultural sector. Since there is no minimum wage in the agricultural sector, the expected wage is equal to the marginal value product of labor determined at full employment of N_A workers. Unlike the starting point in the Lewis model, agricultural wages are assumed to be greater than subsistence level and increasing as workers leave the agricultural workforce. Let agricultural production be represented by the function

$$X_A = q(N_A, \overline{K_A}),$$

where $q' > 0$ and $q'' < 0$, N_A is the workforce in the agricultural sector, and $\overline{K_A}$ is the fixed amount of capital in the agricultural sector, including land.

Now that we have an expression for agricultural production, we can easily derive the marginal product of labor. To find the agricultural wage, we also need to know the price for agricultural output. Since there are only two goods in the economy, manufacturing, and agricultural, we can express the price of the agricultural good according to the terms of trade between the two goods. We can set the price of the manufactured good equal to 1 (i.e., the manufactured good is the numeraire good). Then the price of agricultural output is determined by the relative output of manufactured and agricultural goods. If there is an abundance of agricultural output relative to manufacturing output, then the value of agricultural goods will decrease relative to manufactured goods. Alternatively, if there is a scarcity of agricultural output relative to manufacturing output, then the price of the agricultural goods will rise relative to manufactured goods. We can thus express the relative price of the agricultural good, or terms of trade, with the equation:

$$p = \rho\left(\frac{X_M}{X_A}\right),$$

where $\rho' > 0$.

This expression implies that the price of the agricultural good increases as the relative output of agricultural production to manufacturing production declines. Consequently, as workers migrate from the rural to urban sector, holding all else constant, production in the agricultural sector declines relative to manufacturing, and the price of agricultural products must rise.

Now that we have an expression for agricultural production and for the price of the agricultural good, we can derive an expression for agricultural wages as the marginal value product of labor. That is,

$$w_A = p\frac{\partial q\left(N_A, \overline{K_A}\right)}{\partial N_A}.$$

As rural-urban migration increases, agricultural wages rise for two reasons. First, reduced agricultural production relative to manufacturing production causes the price of agricultural output to rise. Second, the marginal product of agricultural labor rises due to the concavity of the labor production curve.

In equilibrium, the expected wage in the urban sector must be equal to the expected wage in the rural sector. Consequently, the equilibrium condition can be expressed as

$$w_U^E = w_A$$

or, equivalently,

$$\overline{w_M}\frac{N_M}{N_U} = \rho\left(\frac{X_M}{X_A}\right)\frac{\partial q\left(N_A, \overline{K_A}\right)}{\partial N_A}$$

It is clear from this condition that the equilibrating force in this model is migration, since manufacturing wages are fixed, and the size of the labor force in agriculture largely determines agricultural wages.

As an example of how migration maintains equilibrium in the system, suppose the government increases the urban minimum wage. The direct effects of this policy are twofold. First, there is a rise in the expected urban wage. Second, there is a decrease in the quantity of manufactured goods produced, because of a decrease in N_M (since production occurs where the wage equals the marginal value product of labor). This decreases the agricultural terms of trade, $\frac{X_M}{X_A}$, and reduces the price, p, of agricultural goods. These effects cause the left-hand side of the equilibrium equation to rise and the right-hand-side to decrease.

$$\uparrow \overline{w_M} \rightarrow \downarrow X_M \text{ and } \downarrow \rho\left(\frac{X_M}{X_A}\right) \rightarrow \uparrow \overline{w_M} \frac{N_M}{N_U} \text{ and } \downarrow \rho\left(\frac{X_M}{X_A}\right) \frac{\partial q\left(N_A, \overline{K_A}\right)}{\partial N_A}$$

How does the system revert back to an equilibrium state? The answer is that individuals migrate to the urban sector. Increased rural-urban migration (i.e., an increase in N_U accompanied by a decrease in N_A) reduces the expected urban wage by reducing $\frac{N_M}{N_U}$, the probability of obtaining an urban job. It simultaneously increases the agricultural wage by increasing $\frac{\partial q\left(N_A, \overline{K_A}\right)}{\partial N_A}$, due to the concavity of the agricultural production function. Complementing this effect, reduced labor in the agricultural sector reduces agricultural production thereby raising the terms of trade, $\frac{X_M}{X_A}$, and consequently the price of the agricultural good. Rural-urban migration will continue until equilibrium is restored.

REFERENCES

Charlton, D., Taylor, J.E., 2016. A declining farm workforce: analysis of panel data from rural Mexico. Am. J. Agric. Econ. 98 (4), 1158–1180.

Davis, B., Stecklov, G., Winters, P., 2002. Domestic and international migration from rural Mexico: disaggregating the effects of network structure and composition. Popul. Stud. 56 (3), 291–309.

Duflo, E., 2001. Schooling and labor market consequences of school construction in Indonesia: evidence from an unusual policy experiment. Am. Econ. Rev. 91 (4), 795–813.

Farmworker Justice, 2018. Who Works the Fields? The Stories of Americans Who Feed Us, https://www.farmworkerjustice.org/sites/default/files/WhoWorkstheFields-FJ%20Report%20June%202013.pdf. Accessed 14 November 2018.

Guilmoto, C.Z., Sandron, F., 2001. The international dynamics of migration networks in developing countries. In: Population: An English Selection.vol. 13(2). pp. 135–164.

Harris, J.R., Todaro, M.P., 1970. Migration, unemployment and development: a two-sector analysis. Am. Econ. Rev. 60 (1), 126–142.

Lewis, W.A., 1954. Economic development with unlimited supplies of labour. Manch. Sch. 22 (2), 139–191.

Massey, D.S., Arango, J., Hugo, G., Kouaouci, A., Pellegrino, A., Taylor, J.E., 1993. Theories of international migration: A review and appraisal. Popul. Dev. Rev. 431–466.

Mora, J., Taylor, J.E., 2006. Determinants of migration, destination, and sector choice: disentangling individual, household, and community effects. In: International Migration, Remittances, and the Brain Drain, pp. 21–52.

Richter, S., Taylor, J.E., 2007. Gender and the determinants of international migration from rural Mexico over time. In: The International Migration of Women, pp. 51–99.

Chapter 4

Equilibrium and Immigration in the Farm Labor Market

The one constant is that no matter how much we pay, domestic workers are not applying for these jobs. Raising wages only serves to cannibalize from the existing workforce; it does nothing to add new laborers to the pool.

Jason Resnick, VP of Western Growers' Association

There is no such thing as "the farm labor market." Instead, there are many different localized farm labor markets, in which the interaction of labor supply and demand plays out differently in different seasons. Farm labor migration connects these localized labor markets with each other and transmits shocks from one labor market to another. In high-income countries, immigration is the main source of hired workers for many farms, and follow-the-crop migration is the key to redistributing this labor across local labor markets.

In a typical year, between 800 and 900 workers pick the Lake County, California, pear harvest (Rural Migration News, 2007). In 2006, the Lake County pear harvest was the focus of media attention. The *New York Times* ran a front-page story on September 22, 2006, showing pears rotting on the ground, allegedly for lack of labor. A year later, the *Lake County Record Bee* reported that there was an ample supply of pickers for a bountiful pear crop.

What explains such a contrast in the labor situation between only 2 years? The Sacramento delta, to the south, has a larger pear crop than Lake County. In 2006, the Sacramento delta's harvest was late. Migrant workers stayed on to pick in Sacramento, so fewer of them migrated north to Lake County. A labor contractor who brings crews to Lake County explained that he fell behind in harvests near Sacramento and arrived weeks late in Lake County. "There was a lot of pressure on the contractors," he said. "But there is only so much we can do. There wasn't enough labor" (Preston, 2006).

The Farm Labor Problem. https://doi.org/10.1016/B978-0-12-816409-9.00004-5

The Lake County debacle illustrates how an extended demand for workers in one region can create labor shortages in another—even if there are enough workers to pick the crop in a normal year.

This chapter combines principles of farm labor supply and demand to show how interconnected local labor markets interact and reach equilibria that are both seasonal and spatial. They are seasonal because, as we have seen, the demand for farm labor in a given crop tends to be concentrated in only a few months of the year, usually around harvest. They are spatial because harvests happen at different times of year for different crops and in different regions.

In high-income countries, farmers rely heavily on migrant workers to appear at the farm just when needed to harvest the crop, and many workers rely on follow-the-crop migration to piece together employment throughout the course of the year. Farm workers' willingness to engage in follow-the-crop migration is critical to the functioning of local farm labor markets. Without immigration and follow-the-crop migration, farmers would have to rely on local workers, offering them employment for only a few months out of the year.

The interaction of labor supply and demand determines the equilibrium wage and employment in a given farm labor market at a given time of year. Follow-the-crop migration helps to spread the total labor supply across crops and space.

Where does this total labor supply come from? Historically in the United States and other high-income countries, and currently in most developing countries, the majority of hired farm workers have been domestic. However, as incomes rise, domestic workers shift out of farm work into a rapidly expanding manufacturing and service economy. To a large degree, the shift of labor out of agriculture is intergenerational. Parents stay on the farm, while their children, who have access to better economic prospects in the nonfarm economy, move out.

Today, the majority of hired workers on US farms are immigrants. We know this because the US government gathers information about the hired farm workforce and its composition through the National Agricultural Worker Survey (NAWS).[1] Reliable data on the composition of farm workforces in other high-income countries are not so easy to come by, but there is no question that reliance on foreign-born farmworkers is pervasive, not only in rich countries but also in many not-so-rich ones. An abundant, elastic supply of people in poorer countries willing to migrate across borders then work as "follow-the-crop migrants" makes the farm labor supply in rich countries more flexible, so that farmers can meet their seasonal labor demands most of the time. However, unexpected changes in the size or timing of a crop harvest can lead to drastic local labor shortages, as illustrated by the pears rotting on the ground in Lake County in 2006.

---

1. The NAWS website is: https://www.doleta.gov/naws/. Here you can find a wealth of findings from past NAWS survey rounds. Accessed April 24, 2018.

This chapter is about equilibria in interconnected farm labor markets; what happens to these equilibria as domestic workers leave agriculture; and how access to an abundant supply of immigrant agricultural workers creates new equilibria, with high employment and lower wages.

A SPATIAL EQUILIBRIUM

Fig. 4.1 depicts equilibrium in a farm labor market at a particular place and time. Let's call this "Labor Market A." It might represent, for example, the farm labor market in the Sacramento Valley (SV) at the time of the pear harvest. LS_{A0} represents the farm labor supply in this market and LD_{A0} represents the farm labor demand. The equilibrium labor quantity and farm wage are given by the intersection of these two curves, at point a. They are L_{A0} and w_{A0}, respectively.

Over time, as this economy grows and new employment opportunities open up in the nonfarm sector (Chapter 3), the labor supply curve shifts inward and becomes less elastic. This results in a new equilibrium characterized by a higher agricultural wage and lower quantity of labor, as illustrated by point b in Fig. 4.2.

Farm labor migration links the SV farm labor market with the San Joaquin Valley (SJV) to the south, which we can call "Labor Market B." Fig. 4.3 shows a map of the Central Valley of California. The SJV and the SV are contiguous.

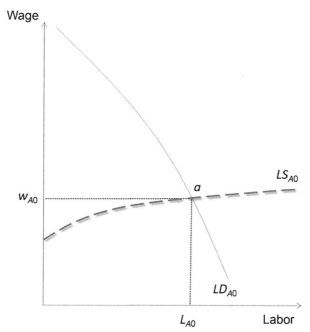

FIG. 4.1 The intersection of supply and demand at point a gives a local equilibrium wage and quantity of labor equal to L_{A0} and w_{A0}, respectively, in locality A in a given season.

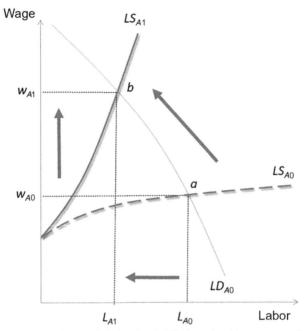

FIG. 4.2   Over time the labor supply in locality A shifts inward and becomes more inelastic (vertical). As a result, the local labor market equilibrium shifts to point b, with a higher wage and lower labor quantity.

Suppose the SJV farm labor market can be represented at this time of year by the labor supply and demand curves in Fig. 4.4. At point c in this figure, there is a lower equilibrium wage and higher equilibrium labor demand than in the SV.

What happens when farm labor migration links these two labor markets? The result is shown in the two graphs in Fig. 4.5. Workers in the SJV will see the higher wage in the SV and, if the cost of migrating is not too high, they will move northward, as described by the mover-stayer model we learned about in Chapter 3. This shifts the SJV labor supply curve inward, from LS_{B1} to LS_{B2}, driving up the SJV equilibrium wage and decreasing the equilibrium labor quantity (point d in left graph). As migrant workers arrive in the SV, the SV labor supply curve shifts outward, to LS_{A2} (right graph). Access to the SJV workers reduces the equilibrium wage and increases the labor quantity in the SV (point e in right graph). If migration costs are low enough, the wage gap between the two regions approaches zero, as illustrated in Fig. 4.5 (the wage is the same at points d and e). Otherwise, the difference in wages between the two labor markets reflects migration costs and people's willingness to migrate to farm jobs.

If migration costs are negligible, we can combine the graphs in Fig. 4.5 into one diagram representing a single regional labor market. In Fig. 4.6, the bold

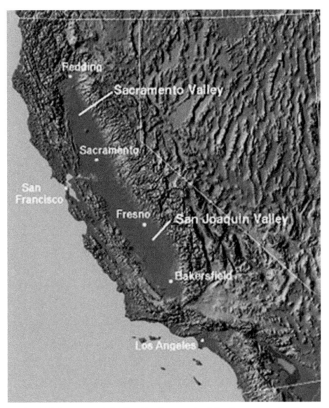

FIG. 4.3 California's San Joaquin Valley lies just south of the Sacramento Valley. The two constitute the Central Valley. *(From http://centroculturalaustriaco.com/where-is-california-city-on-the-map/central-valley-california-california-road-map-where-is-california-city-on-the-map/.)*

supply and demand curves are the horizontal summations of the two localities' supply and demand. The intersection of the regional supply and demand curves yields the regional equilibrium wage and employment level. (Note that the equilibrium wage is the same here as in Fig. 4.5.)

The new spatial equilibrium (point *f*) results in a shared wage, w_{AB}, between the two regions, as migrant workers move between labor markets, redistributing employment from south to north.

In the next season, the SV harvest winds down and perhaps a new harvest of navel oranges begins in the SJV, so the migrant workers might move south again. Alternatively, if a new harvest opens up farther to the north, say, in Oregon wine grapes or Washington state apples, workers might continue heading northward once the Sacramento pears are harvested.

We could draw many different figures depicting interlinked farm labor markets all around the United States and other countries. The figures would all shift

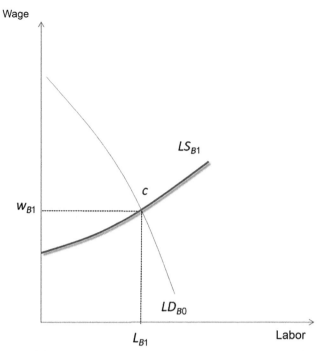

FIG. 4.4 The equilibrium wage is lower in a neighboring labor market (locality B).

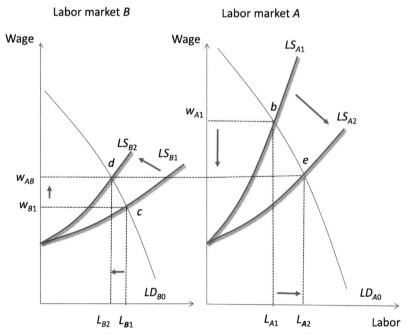

FIG. 4.5 Farm labor migration links the two local labor markets, resulting in a new regional equilibrium wage.

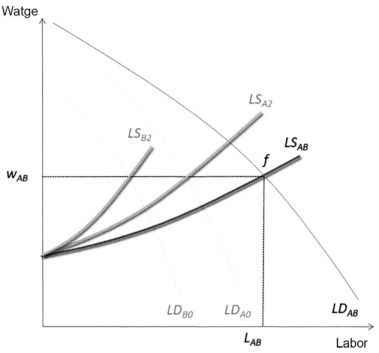

FIG. 4.6   The two local labor markets merge into a single regional labor market when labor is mobile.

around from season to season, as harvests of some commodities in some regions begin and harvests of commodities in other regions wrap up. Gravity makes water flow from high places to low places. Expected earnings gaps induce workers to flow from places where expected earnings are low to places where they are high.

IMMIGRATION CHANGES EVERYTHING

Imagine a reservoir from which water flows out to farmers' fields. As one field starts to irrigate, the farmer opens its canal and the water flows. After she finishes irrigating, she closes the canal, and the water flows to other fields. As long as there is water in the reservoir, fields will be irrigated, and water will flow all across the agricultural landscape. As one farm's water demand is met, water becomes available to other farms.

Something similar happens with farm labor when there is a plentiful supply of foreign workers. Follow-the-crop migration redistributes workers across local labor markets in response to expected earnings differences. Immigration redistributes workers across national boundaries in response to expected earnings differences. Fig. 4.7 illustrates the case in which the supply of foreign labor

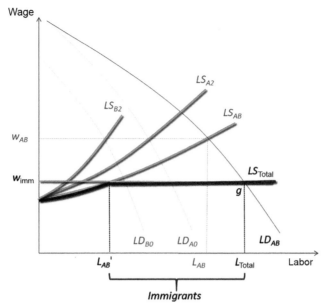

FIG. 4.7 With internal migration and a perfectly elastic supply of foreign workers, wages in all labor markets are replaced by the immigrants' reservation wage—similar to the Lewis model.

is perfectly elastic (flat), like in a Lewis model. The foreign workers' reservation wage is w_{imm}. This is the minimum that farmers have to offer in order to tip the foreign workers' mover-stayer model in favor of migrating across the border. It has to equal at least the workers' expected earnings at home *plus* migration costs.

Most immigrant farmworkers in the United States and many or most in other high-income countries do not have legal immigration status. Some countries have guest worker programs, under which a specific number of farm workers from low-income countries can enter seasonally to do farm work, as we shall see in Chapter 6. Either way, the cost of the border crossing (and in some migration systems, a sea crossing) is likely to include fees migrants pay to labor recruiters or human smugglers. Intensified border enforcement increases human smuggler fees. International labor recruiters may be able to charge migrants more for their services when governments limit the number of guest worker permits that are available and many workers in migrant-sending countries vie for a small number of legal jobs abroad. When workers have to pay more to migrate across borders, the reservation wage rises.

The reservation wage plays the same role in our model as the subsistence wage does in a Lewis model. In fact, it sets the wage for both foreign and domestic workers and (adjusting for migration costs within countries) for all local farm labor markets.

In Fig. 4.7, access to foreign agricultural workers lowers the equilibrium wage to w_{imm} in both labor market A and B. At the equilibrium (point g), employment is L_{total}, of which L_{AB}' is domestic workers and $L_{total}-L_{AB}'$ is immigrant workers.

When immigration, coupled with internal migration, links farm labor markets in high-income countries to foreign labor markets, the domestic labor supply can become largely irrelevant to farm labor market outcomes. To see this, suppose that domestic workers continue moving out of farm work in both local markets, but immigrant workers are available to fill the resulting void in the farm labor supply. The supply curve for domestic workers shifts inward and becomes steeper, as shown in Fig. 4.8. However, the equilibrium wage and employment do not change; the equilibrium remains at point g. There is simply a higher share of immigrant workers in the farm workforce.

The share of immigrants in the workforce depicted in Fig. 4.8 might seem large. However, California's farm labor markets are not much different than in this figure; fewer than 1 in 10 hired farmworkers there are US born. The rest are foreign-born, overwhelmingly from rural Mexico.

The dominant role of immigrants in the hired farm work force is controversial almost everywhere in the world. Even in the richest countries, some people argue that the small share of domestic workers on farms is the result of immigrant workers undercutting domestic worker wages and driving domestic

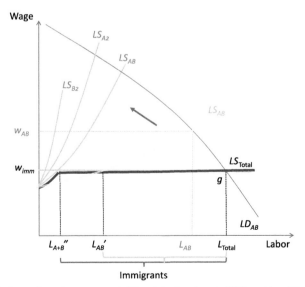

FIG. 4.8 As domestic workers move out of farm jobs, immigrants fill the void, and the equilibrium wage does not rise.

workers out of farm jobs. It could also result from immigrants doing farm work that native workers avoid, consistent with an inward shifting, inelastic supply of domestic labor, like in Fig. 4.8.

This is a chicken-and-egg question. Would the farm labor supply shift inward with or without immigration? Notoriously difficult working conditions, characteristic of farm work, might discourage native workers with other options from working in the fields. There is no definitive evidence proving that immigrants do or do not drive domestic workers out of the fields, because we will never know what would have happened without access to foreign agricultural workers. However, there is some anecdotal and "quasiexperimental" evidence that workers in high-income countries would not be interested in doing farm work even if wages were higher than they are now. We discuss this evidence further in Chapter 6.

Thinking about this question more deeply, what would happen if there were no farm labor immigration? Rich countries would grow less food, and the food would be more expensive. That would be bad for consumers, and it would reduce revenue from food exports. Farm wages would be high, which might be good for farm workers.

On the other hand, there would be intense pressure to find substitutes for scarce, expensive farm workers. The static view that food prices and farm wages would rise ignores the dynamic adjustments that undoubtedly would occur. Governments, universities, and entrepreneurs would have incentives to develop new technologies like machines to harvest fruits and vegetables. Farmers would have profit motives to invest in these technologies.

Workers would have to adapt, too. A mechanical harvester does not require low-skilled workers like the ones that harvest most fruit and vegetable crops today. It requires workers who understand the technology and have the skills to operate and maintain the machinery. Would you be willing to do that for $20 an hour? Or work in a startup company as an engineer inventing a new machine to pick peaches or strawberries? The fields would look different without low-cost foreign labor—and so would the workers.

FOLLOW-THE-CROP MIGRATION REDISTRIBUTES WORKERS WITHIN COUNTRIES

We have seen that the total farm labor supply can draw from both domestic and foreign households, and the immigrant share of the farm workforce typically increases as countries become richer and less willing to supply their own labor to farm work. Follow-the-crop migration is the mechanism that redistributes the total farm labor supply across space to ensure equilibrium across local farm labor markets. Without follow-the-crop migration, it would not be possible to have a single regional farm labor market like in Figs. 4.5 and 4.6. Instead, labor would become trapped in local farm labor markets like in Figs. 4.2 and 4.3.

Macroeconomics teaches us that if money moves quickly through an economy (i.e., the *velocity of money* is high), a smaller total money supply will be needed to support a country's economic activity. Something similar happens with farm labor. If workers move quickly from farm to farm (i.e., *follow-the-crop migration* is high), a smaller total farm labor supply is needed to bring all the crops in. Without follow-the-crop migration, many more workers would be required to make sure that there is enough labor at each location and time to harvest crops. Researchers use mover-stayer models to model workers' decisions to engage in follow-the-crop migration.

We can adapt the mover-stayer model in Chapter 3 to study follow-the-crop migration. Wages and labor demands shift continuously across the seasons and across space, sometimes literally from 1 day to the next. At time t, person i's expected income from staying where she is to do farm work is Y_{0it}, and her expected income from engaging in follow-the-crop migration is Y_{1it}. There is a migration cost, which could be psychological as well as monetary, and it also may vary across individuals, space, and time. For example, it might be difficult for a woman to engage in long-distance follow-the-crop migration when her children are out of school and there is no one to care for them. A high cost can make migration unresponsive to expected earnings differences across space, or inelastic. We can denote this cost by C_{it}. The expected economic returns and costs from migrating depend on the person's characteristics, X_i. Is the person male or female, high or low education, married with children or single and flexible, young or old? Let D_{it} be a dichotomous variable equal to 1 if person i engages in follow-the-crop migration at time t and zero otherwise. Then, the probability of migrating, $Pr(D_{it} = 1)$, is simply the probability that the net economic gains from migrating exceed the costs:

$$Pr(D_{it} = 1) = Pr[Y_{1it}(X_i) - Y_{0it}(X_i) - C_{it}(X_i) + \varepsilon_{it}] > 0$$

Fan et al. (2015) used data from the NAWS to estimate a follow-the-crop migration model similar to the one above. They found that people's income and sociodemographic characteristics are the biggest drivers of follow-the-crop migration. People with higher income where they currently are migrate less. People are also significantly less likely to engage in follow-the-crop migration if they are women, unauthorized immigrants, English proficient, married, or have children. African Americans are significantly less likely to migrate than other ethnic groups.

Follow-the-crop migration is an essential feature of commercial agricultural systems in high-income countries, particularly where there is seasonal production of fruits, vegetables, and horticultural crops. However, as we shall see in Chapter 8, agricultural workers are becoming less and less willing to migrate, just as they are becoming less willing to do farm work. Migration decreases as incomes rise and "settling down" becomes more of an option. In the United States, the share of agricultural workers who migrate has fallen by more than 60% since the late 1990s.

LINKING LABOR SUPPLY WITH DEMAND: FARM WORKER RECRUITMENT

What connects the demand for labor on large numbers of farms with the supply of labor by thousands or hundreds of thousands of workers? If you were a farmer, where would you find enough workers at the moment you need them to harvest your crop? And if you were a worker, how would you find out which farms require your services on a particular day? Markets link the supply and demand for everything in an economy, from apples and oranges to real estate, credit, and insurance. Labor is no different, and a variety of institutions have emerged to match workers with farm jobs. Some have proven to be more successful than others.

Direct Hiring by Farmers

Most farmers hire at least some of their workers directly. Typically, farmers rely on their labor supervisors to recruit new workers. Labor supervisors, in turn, tap networks of existing farm workers' friends and family members. An advantage of this strategy is that it can dramatically lower the transaction costs of discovering and recruiting new workers—word of mouth takes the place of the search costs supervisors and farmers would have to incur without access to informal networks. Another advantage of network recruitment is that current workers might have an incentive to refer hard-working family members and friends to become their future coworkers.

Some farmers get to know their workers' families, and they may even provide support to the communities from which their workers come. Investments in people, like these, can win workers' allegiance and commitment to return for the harvest each year. This strategy obviously is more difficult the larger the seasonal workforce a farm employs. Securing a sufficient number of harvest workers at the time and place needed can prove to be an insurmountable challenge to a farmer with large seasonal labor demands.

Farmer Associations

Sometimes groups of farmers form associations that recruit and share workers. A well-known example in the 1970s was the Coastal Growers Association (CGA) of lemon producers in Ventura, California, which offered wages, fringe benefits, and job stability. By holding onto productive workers, the CGA managed to raise workers' earnings without significantly increasing harvest costs. Workers returned annually or settled in the area. However, this labor management model proved unsustainable. The CGA was not able to secure a replacement workforce as its workers aged or moved out of farm jobs, because new workers were difficult to recruit and the farmworkers' children rejected harvesting jobs. Eventually, the CGA farmers turned to farm labor contractors (FLCs) to meet their harvest needs (Martin, 2003; Lloyd et al., 1988).

Unions and Hiring Halls

In theory, farmworker unions could act as hiring halls, allocating their members across farms as needed to harvest crops or perform other seasonal tasks. In fact, union contracts could offer farmers the assurance of having timely access to workers while at the same time providing workers with benefits in the form of better wages and working conditions. However, very few farm workers in the world today are members of unions, even in California where Cesar Chavez and the United Farmworkers Union made history by organizing large numbers of farm workers in the 1960s and 1970s. Unions' efforts to operate as farmworker hiring halls largely have failed.

Public Employment Services

In some countries, public employment services (PESs) offer support to unemployed workers, including unemployment insurance and assistance finding jobs. In the United Kingdom, for example, employers can advertise job vacancies and jobseekers can look for jobs through a computer system called Universal Jobmatch. On April 24, 2018, a universal jobmatch search for "farm worker" jobs across the entire United Kingdom produced 22 results.[2]

The US Department of Labor offers assistance in finding jobs through its CareerOneStop website. A search on "farm worker" in California produced 55 results, including "farmworker/laborer/crop" jobs but also more skilled "agricultural equipment operators" and even several "emergency medical technician" jobs at chicken processing plants.[3]

Farmers often complain that PES cannot provide them with the numbers of committed farm workers that they require. Lawsuits in the United States have accused PES in agricultural areas of discriminating against farm workers by not telling them about nonfarm jobs—in effect putting PES in the position of removing workers from agriculture. Another explanation for PES' generally dismal performance connecting workers with farm jobs concerns the workers, themselves. One study concluded: "It appears that almost all US workers prefer almost any labor-market outcome—including long periods of unemployment—to carrying out manual harvest and planting labor" (Clemens, 2013).

---

2. https://jobsearch.direct.gov.uk/JobSearch/, Accessed 24 April 2018, search on "farm worker," no specified skills, in "United Kingdom," posted in the last 30 days.
3. http://www.careeronestop.org/Toolkit/Jobs/find-jobs.aspx?keyword=farm%20worker& ajax=0&location=california&source=DEA, Accessed 24 April 2018, search on "farm worker," "California."

Farm Labor Contractors

The seasonality of farm labor has led to the creation of a specialized institution in the farm labor market: FLCs. FLCs specialize in finding workers and supplying them to farms, for a fee. Their comparative advantages are having access to networks of contacts through which they can recruit large numbers of workers, and their ability to quickly move workers from one farm to another. Instead of hiring workers directly, a farmer may enter into a contract with a FLC to harvest the crop at an agreed-upon price. This price, in theory, should cover the wages the FLC will have to pay his workers, any mandatory payroll taxes and other costs, like transportation, as well as the FLC's profit.

FLCs play an important and growing role in matching workers with seasonal farm work around the world. In a mover-stayer model, they lower the costs to workers of finding farm jobs. They also lower the costs to farmers of finding workers.

In practice, there are few barriers to entry and many FLCs, so FLCs have an incentive to undercut one another in order to win farmer contracts, and the prices they charge farmers often are not sufficient to cover the minimum wage plus government-mandated payroll taxes (if applicable), while turning a profit. Unscrupulous FLCs fill the gap by charging vulnerable workers for mandatory rides to the fields, meals, and other services; paying below the minimum wage; and evading taxes and labor and immigration laws (Martin and Taylor, 2000).

FLCs know their workers well. In California, many FLCs previously were farm workers. Most of their workers are unauthorized immigrants with a fear of being deported, little education, and limited understanding of their labor rights. A FLC can exploit workers who perceive that they are in a weak bargaining position. Of course, there are many reputable FLCs who comply with labor laws, but they find themselves at a disadvantage when others undercut them by exploiting their workers. This is a classic case in which free markets do not lead to the greater good. California's Agricultural Labor Relations Board (ALRB) was created by law in part to deal with these sorts of situations.

Reliable information about the activities of FLCs is hard to come by in the United States, and even more so in other countries.

Farm Labor Recruiting in the Internet Age

Not all job matchers fall into neat categories like the ones above. The WAFLA, which started out being a Washington-state Farm Labor Association, seeks to provide seasonal farm employers in the Pacific Northwest of the United States with a stable and legal farm workforce. It makes extensive use of information technology to provide member farmers with advice and labor and employment consulting services in "attracting a quality workforce." The WAFLA also hires workers, including H-2A agricultural guest workers (see Chapter 6), thereby performing many of the functions of a large-scale FLC that moves crews from farm to farm.[4]

A number of online agencies link potential workers with agricultural employers but without physically moving workers across farms like FLCs do. One example, Picking Jobs!, matches people interested in doing seasonal farm work (including students) with agricultural employers, primarily in the European Union countries.[5] In theory, the internet could reduce the transaction costs of matching large numbers of workers with seasonal farm jobs dramatically, but in practice its impact has yet to materialize on a large scale.

THE CHALLENGES OF SEASONALITY FOR WORKERS AND COMMUNITIES

From a worker's point of view, the seasonality of farm labor demand means that wages are a poor reflection of actual earnings. A fast picker earning piece rate can make an hourly wage well above the minimum agricultural wage, but only when there is a crop to pick. He may show up at a site where FLCs recruit workers early in the morning, only to find that his services are not needed that day. Both the seasonality of employment and low wages explain why most farmworker households around the world have income well below the poverty line.

Rural communities everywhere have to deal with the needs of large numbers of seasonally employed, impoverished farm workers and their families. An industry with highly seasonal labor demands is not conducive to maintaining stable, prosperous communities. In fact, a study by Martin and Taylor found that an additional nonfarm job in the United States reduces poverty, but an additional farm job increases it (Martin and Taylor, 2003).

IMMIGRATION AND FARM LABOR ACROSS COUNTRIES

No other country offers anywhere near the same richness of econometric and policy studies—not to mention hard data—on farm labor migration as the United States does. Nevertheless, cross-country analysis of migration policies and anecdotal evidence on farm labor and immigration trends highlight the universality of the farm labor problem. Almost without exception, high-income countries depend heavily on immigrant agricultural workers—and many not-so-high-income countries do, too. Here are some examples:

- New Zealand imports guest workers from Pacific Island nations, primarily to harvest wine grapes, kiwifruit, and apples—half of which are exported. Most immigrant farmworkers are from Vanuatu and Tonga.

---

4. See Wafla Mission, https://www.wafla.org/mission, Accessed 24 April 2018.
5. See www.pickingjobs.com.

- Canada brings seasonal farm workers from the Caribbean (particularly Jamaica) and Mexico, and there appears to be an increasing number of unauthorized Mexican immigrants working on Canadian farms.
- Spain, a major supplier of fresh fruits and vegetables to the European Union, employs what might be the most diverse farm workforce among high-income countries, with immigrants from Morocco, Latin America, Romania, and other lower-income countries.[6]
- Greek agriculture relies significantly on Albanian workers. Many Albanians entered as illegal migrants in the 1990s, following the collapse of Albania's communist government in 1991 and a financial crisis provoked by pyramid schemes in the banking sector in 1996.
- The United Kingdom imports farmworkers from the A8 Countries of Eastern Europe, especially Poland. After Britain announced its decision to leave the European Union (Brexit), Britain faced severe labor shortages. There is considerable uncertainty about how Britain's agricultural sector will adapt to Brexit, inasmuch as the majority of Britain's current farm workers are Eastern Europeans who will no longer have seamless access to UK farms following Brexit. The United Kingdom is currently considering implementing an agricultural guest worker program.
- South African farmers recruit farm workers from nearby Zimbabwe.
- Haitians constitute over 90% of the seasonal sugar work force and two-thirds of coffee workers in the Dominican Republic. At least 32% of total agricultural wages in the country are paid to Haitians (Martin et al., 2002; Filipski et al., 2011).
- Costa Rica has a long history of importing farmworkers from its poorer Central American neighbor, Nicaragua. There were an estimated 135,579 non-naturalized Nicaraguans in Costa Rica in 2000, up from 78,487 in 1998. One in four Nicaraguan immigrant workers in Costa Rica works in agriculture (International Organization for Migration, 2001).
- Sri Lanka is a major exporter of domestic helpers and other workers; however, it imports Indian workers to harvest tea.
- Most workers employed on Malaysian palm oil plantations are Indonesians and Bangladeshis (Martin, 2016).
- Mexico is unique in being a major country of farm labor emigration, internal migration, and immigration. It is by far the world's largest nation of farm labor emigration, supplying most of the US hired farm work force. Significant internal migration supplies labor to agribusinesses in Mexico, largely producers for export markets in the northwestern states of Sinaloa, Sonora, Baja California, and Baja California Sur. The migrant workers are mostly from Mexico's poorer southern states of Oaxaca and Guerrero. Mexico also imports agricultural workers from Guatemala to fill low-wage jobs in the South.

---

6. For more information about foreign agricultural workers in Canada, New Zealand, Australia, Spain, and the United States see P.L. Martin (2016).

APPENDIX A

Modeling Farm Labor Market Equilibrium

Farm labor market equilibrium is found where the quantity of farm workers demanded by farms equals the quantity willing to supply their labor to agriculture. In the chapter on farm labor demand, we saw that farmers hire workers at the point where the marginal value product of labor is equal to the wage. Given diminishing marginal returns to labor, the demand for farm workers must be decreasing in wages. In contrast, the farm labor supply is increasing in wages, since higher wages encourage people to work in agriculture. Its slope depends on the elasticity of supply, or how willing domestic and/or foreign workers are to supply their labor to farm work.

We can write a system of two equations for demand and supply and solve the system for the equilibrium wage and labor employed. Begin with a model with no foreign workers. Let the labor demand in a market closed to immigration be

$$LD_0 = \alpha_0 - \alpha_1 w_0$$

where LD_0 is the quantity of farm labor demanded in a market with only domestic workers, $\alpha_0 > 0$, $\alpha_1 > 0$, and w_0 is the equilibrium wage.

The domestic labor supply is expressed as

$$LS_{dom} = \beta_0 + \beta_1 w_0$$

where LS_{dom} is the quantity of domestic labor supplied, $\beta_0 > 0$, and $\beta_1 > 0$.

In equilibrium, markets must clear, so the quantity of labor demanded is equal to the quantity of labor supplied: $LD_0 = LS_{dom}$. This implies

$$\alpha_0 - \alpha_1 w_0 = \beta_0 + \beta_1 w_0$$

By solving this equality for w_0, we can derive an equation for the equilibrium farm wage without immigration, as a function of the parameters in the farm labor demand and supply functions:

$$w_0 = \frac{\alpha_0 - \beta_0}{\alpha_1 + \beta_1}$$

We can substitute this expression for the wage back into the labor supply and demand equations to derive an equation for the equilibrium quantity of labor demanded and supplied:

$$LD_0 = LS_{dom} = \alpha_0 - \alpha_1 \left(\frac{\alpha_0 - \beta_0}{\alpha_1 + \beta_1} \right)$$

With these two equations, we know the equilibrium farm wage and employment once we know or estimate the parameters in the farm labor demand and supply functions.

Now consider what happens to market wages, labor demand, and domestic labor supply when the market opens up to immigration. Suppose there is an infinitely elastic supply of foreign workers willing to work at wage $w_{imm} < w_0$. The labor demand in a market with immigration is then

$$LD_1 = \alpha_0 - \alpha_1 w_{imm}$$

It is clear that $LD_1 > LD_0$ if $w_{imm} < w_0$. That is, the number of farm workers demanded in a market with access to immigrant workers willing to do farm work at a low wage is greater that the number demanded in a high-wage market closed to immigration.

How many domestic workers are willing to work at the new market-clearing wage?

$$LS_{dom}' = \beta_0 + \beta_1 w_{imm}$$

Since $w_{imm} < w_0$, it is clear that $LS_{dom}' < LS_{dom}$. That is, the number of domestic workers willing to work in agriculture in a market open to low-wage immigrant workers is less than the number of domestic workers supplying labor to agriculture in the closed, high-wage economy.

The difference between the quantity of workers demanded and the domestic worker supply, $LD_1 - LS_{dom}'$, is the number of immigrants working in the agricultural sector. The more sensitive the domestic worker supply response is to wages (i.e., the larger β_1 is), the more immigrants will take the place of domestic workers in the farm labor market. Mathematically, if β_1 is large, the drop in wage from w_0 to w_{imm} will provoke a large number of domestic workers to leave farm jobs.

REFERENCES

Clemens, M., 2013. The Effect of Foreign Labor on Native Employment A Job-Specific Approach and Application to North Carolina Farms. Center for Global Development Working Paper No. 326 (May), p.25. https://www.cgdev.org/sites/default/files/effect-foreign-labor-native-employment.pdf.

Fan, M., Gabbard, S., Alves Pena, A., Perloff, J.M., 2015. Why do fewer agricultural workers migrate now? Am. J. Agric. Econ. 97 (3), 665–679.

Filipski, M., Taylor, J.E., Msangi, S., 2011. Effects of free trade on women and immigrants: CAFTA and the rural Dominican Republic. World Dev. 39 (10), 1862–1877.

International Organization for Migration, 2001. Binational Study: The State of Migration Flows between Nicaragua and Costa Rica. International Organization for Migration, Geneva. https://publications.iom.int/books/binational-study-state-migration-flows-between-costa-rica-and-nicaragua.

Lloyd, J., Martin, P.L., Mamer, J., 1988. The Ventura Citrus Labor Market. Giannini Information Series No. 88-1, University of California Division of Agriculture and Natural Resources.

Martin, P., 2003. Promise Unfulfilled: Unions, Immigration, and the Farm Workers. Cornell University Press, Ithaca, p. 51.

Martin, P.L., 2016. Migrant Workers in Commercial Agriculture. International Labour Organisation, Geneva.

Martin, P.L., Taylor, J.E., 2003. Farm employment, immigration and poverty: a structural analysis. J. Agric. Resour. Econ. 28 (2), 349–363(August).

Martin, P., Taylor, J.E., 2000. For California farmworkers, future holds little prospect for change. Calif. Agric. 54 (1), 19–25. https://doi.org/10.3733/ca.v054n01p19.

Martin, P.L., Midgley, E., Teitelbaum, M.S., 2002. Migration and development: whither the Dominican Republic and Haiti? International Migration Review 36 (2), 570–592.

Preston, J., 2006. Pickers Are Few, and Growers Blame Congress. New York Times. September 22. http://www.nytimes.com/2006/09/22/washington/22growers.html?n=Top%2FReference%2FTimes%20Topics%2FSubjects%2FF%2FFood&_r=1. Accessed 24 April 2018.

Rural Migration News, Vol. 13, No. 4, 2007.

Chapter 5

Labor in an Agricultural Household

The earth is given as a common stock for man to labour and live on…as few as possible shall be without a little portion of land. The small landholders are the most precious part of a state.

Thomas Jefferson, in letter to James Madison in 1785

Far and away most of the world's agricultural producers are also households, cultivating small-sized landholdings. As producers, they combine inputs, including labour, to produce a harvest. As consumers, they take their income from agricultural production and other sources, including wage labour they supply to other farms and businesses, and use it to purchase food and other items for the household. More often than not, part or all of a household's crop production gets consumed by the household that produced it, and part or all of the labour the household uses to produce crops comes from the household, itself. Households have their own internal labour market equilibrium: their demand for labour on the farm must match their supply of time to work, plus or minus the labour they hire in or out. In the poorest agricultural economies, many households do not have access to hired farm workers or to off-farm wage employment opportunities. For these households, the wage is an endogenous "shadow wage," reflecting the household's own valuation of time, not a wage set exogenously by a market. As agricultural economies evolve, agricultural households become more integrated with outside labour markets. In this chapter, we learn how to use agricultural household models to understand labour demand, supply, and equilibrium in poor traditional economies and the evolution of agricultural labour markets at early stages of countries' economic development.

Today, in the world's poorest countries, most of the workforce is in agriculture, and most agricultural producers are households. The same is true historically of today's richest countries. Thomas Jefferson, one of the US founding fathers, argued that family farms were the foundation of an affluent and virtuous society. The US farm labor market is very different from the American workforce that Thomas Jefferson envisioned in his letter to James Madison in 1785. Was Jefferson's vision ever a reality in the United States? At times and in certain regions of the United States, there have been examples of small farmers living

The Farm Labor Problem. https://doi.org/10.1016/B978-0-12-816409-9.00005-7

and laboring on the land, but that has not been the predominant story of US fruit, vegetable, and horticultural (FVH) production over the past century. Rather, larger farms that employ immigrants to work the land dominate the history of FVH agriculture, particularly in the western United States. In 1850, just over one-half of the entire US population lived on farms. By 1990, only 1% did.[1] Nevertheless, most of the world's 570 million farms are family run, and family farms cultivate about 75% of the world's agricultural land (Lowder et al., 2016).

We begin our journey in this chapter by exploring the role of labor in traditional agricultural systems, in which households are both producers and consumers of crops and supply most of their own crop inputs—including labor. To study the role of agricultural labor in traditional agricultural systems, we need to understand the dual nature of the agricultural household as consumer and producer.

Agricultural household models are the economist's workhorse for studying labor, production, and consumption in rural areas of developing economies. Agricultural households act as both producers and consumers of the same goods. This dual function makes agricultural households distinct from pure firms, which solely maximize profits, and from pure consumers, who maximize their utility or satisfaction by choosing an optimal combination of goods to consume, subject to a budget constraint. In an agricultural household, the household supplies many, even all, of the inputs for production, and it consumes some or all of the food it produces. Labor is a particularly interesting input, because it enters into both the household's production problem, as an input, and its consumption optimization problem, through the work-leisure tradeoff. (In Chapter 3, we considered leisure as a consumer good that makes the household better off.)

Agricultural households implicitly purchase output from their own production, and they purchase leisure from themselves by not working. Households are endowed with a fixed number of hours each period in which they can either work or demand leisure. Each hour spent in leisure could be spent at work, instead. Consequently, an hour of leisure is valued at the marginal earnings of an additional hour of work. The marginal earnings from another hour of work are what the household would have earned if it had worked during that hour instead of enjoying an hour of leisure. We call these foregone earnings the "opportunity cost of leisure."

When households have access to perfect markets, the opportunity cost of leisure is equal to the market wage, and the value of an additional unit of food or other consumption goods equals the market price.

When markets determine prices, the household's production and consumption problems are said to be separable from one another or recursive, meaning that the household's production and consumption can be solved sequentially. In order to maximize utility, the household first maximizes its income by making the optimal decisions on the production side. Once it accomplishes this, the

---

1. Agriculture in the Classroom, "Historical Timeline—Farmers and the Land," https://www.agclassroom.org/gan/timeline/farmers_land.htm.

household can act as a consumer, maximizing utility by allocating its income between consumption of the agricultural good (valued at its market price) and leisure (valued at the market wage). With perfect access to labor and consumption markets, the household's consumption decisions have no influence on its production decisions, and its production decisions only influence its consumption decisions by determining total income and consequently the budget constraint.

High transaction costs often make entering the markets for labor or agricultural goods prohibitive in rural communities. (Transaction costs are the costs of buying and selling. You can learn about them in Taylor and Lybbert, 2015.) Then, households can no longer buy or sell goods or labor at the market price. Most agricultural households in the world do not sell their produce in formal markets—or if they do, they sell only a very small part of the harvest. The transaction costs of getting access to markets are simply too high. On the labor side, even when there is some wage labor employment in a village, this does not mean that a household can work all of the hours that it wishes to work at the market wage, or that it can hire workers who are perfect substitutes for family workers.

When the household does not have access to markets, the nonmarketable goods still have prices, but the prices are invisible, subjective valuations, internal to the household. These prices are equal to the opportunity cost of resources devoted to the production of that good, which could be employed elsewhere. Consequently, they may differ across households. We call these prices "shadow prices." (We call the household's subjective valuation of labor in the absence of labor markets the "shadow wage.")

Shadow prices or shadow wages are determined endogenously by the household's supply and demand for nonmarketed goods or labor. The agricultural household model without access to markets is said to be nonseparable, because the household must make its production and consumption decisions simultaneously, not sequentially. Whatever it demands, it must supply, and vice versa. If a household member wishes to enjoy an additional hour of leisure, the household must reduce its production because it cannot hire an additional hour of labor to keep its production where it was. There is a production-leisure trade-off, and household preferences enter directly into production decisions. Even if households lack access to a market for only one good, consumption and production are affected.

Households are always at least as well off when they can trade goods and labor instead of consuming only what they produce, but trade can also make households susceptible to sudden changes in prices, which may or may not improve the household's well-being. The unique properties of agricultural households as producers and consumers of the same goods often make the impacts of wage and price changes on employment, income, and welfare hard to predict. Outcomes are especially difficult to predict when the market for labor and one or more consumer goods is missing.

We first examine the roles of agricultural households as producers and consumers separately. Then we learn how to model agricultural households' farm labor supply and demand in a traditional economy with no access to markets. As economies grow and develop, markets for goods and labor materialize around agricultural households. We will learn how this impacts households' consumption, production, and labor decisions, as well as their welfare. We perform some *comparative statics* to see how agricultural households respond to price or policy changes in different situations.

AGRICULTURAL HOUSEHOLDS AS PRODUCERS

To start, consider the household's production decision. To keep the model as simple as possible, we'll assume that households are endowed with a fixed quantity of capital and land, which together we denote by $\overline{K}$, and they demand L_D units of labor at a wage w to produce Q_S units of food, where the subscripts D and S represent demand and supply, respectively. A production function $Q_S = Q(L_D, \overline{K})$ describes the technology or production function the household uses to convert labor and capital into food. This production function is monotonically increasing in labor and capital, and it exhibits diminishing marginal returns to factor inputs, the usual assumptions in producer theory.

Let p be the price of food. Household profits are total revenue from producing and selling food minus the variable cost of production, which in this model is simply the cost of labor, wL_D. Profits can then be expressed as $\pi = pQ(L_D, \overline{K}) - wL_D$.

In the basic labor supply model we studied in Chapter 3, the household's potential or *full income* was simply the value of its time endowment: $Y_f = wT$, where T is the household's time endowment, and w is the wage earned per unit of time employed. In an agricultural household model, full income also includes the household's profit from crop production: $Y_f = wT + \pi$. Keep in mind that higher wages increase the value of the household's time endowment, but they reduce household-farm profits, because labor is a cost in the profit function.

Fig. 5.1 illustrates this production function and the household's production decision. The horizontal axis measures units of labor (L), and the vertical axis is the quantity of food produced by the household. Notice that the production function is concave, due to diminishing marginal returns to labor. That is, with each additional unit of labor employed, production increases by a smaller amount. The slope of the production function at any given point is equal to the *marginal rate of transformation* (*MRT*) of labor into food, that is, the number of units of food that the household can produce by employing one more unit of labor. Mathematically, the *MRT* is the derivative of the production function with respect to labor employed, written $\frac{dQ(L_D, \overline{K})}{dL_D}$. You will recognize this as the marginal product of labor (*MPL*) from Chapter 2.

In Fig. 5.1, the household maximizes profits where the slope of the production curve, or *MRT*, equals the ratio of labor wages to the price of food. This point of tangency identifies the optimal production, Q_S^*, and the optimal amount of labor employed, L_D^*.

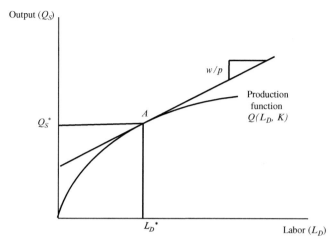

FIG. 5.1 The agricultural household produces food at point A, where the slope of the production curve just equals the ratio of the wage to the price of food.

AGRICULTURAL HOUSEHOLDS AS CONSUMERS

Agricultural households are consumers as well as producers, and we analyze their consumption behavior using consumer theory. As a consumer, the household maximizes utility by allocating L_f^* units of its time to work, so that it can consume X_c^* units of the good that it produces. (Later we shall see what happens when the household can buy consumption goods.) The subscript f indicates the amount of labor the household supplies from the *family's* labor endowment, and the subscript c indicates the amount of food the household *consumes*.

The household's well-being, or utility, is increasing in its food consumption, X_c, and decreasing in its supply of time to work, L_f. Leisure increases utility, and work does the opposite. There is a one-to-one trade-off between leisure and labor supply, so one is the mirror image of the other. The household has a fixed amount of time each period in which it can work or enjoy leisure. Let us call this total time endowment $\overline{T}$ and the leisure that the household takes (or *consumes*) l_c. Then, leisure can be calculated as the difference between the total time endowment and the amount of time the family supplies to work, that is, $l_c = \overline{T} - L_f$.

The household's well-being increases with both the consumption of leisure time and the consumption of food. However, there is a trade-off between consuming these two goods. If the household consumes its entire time endowment in leisure, it will have no time to allocate to work to produce (and therefore consume) food. Alternatively, if the household devotes all of its time to work, it can consume a large quantity of food, but it will enjoy little leisure. It will value a little more leisure time more than a little more food, and that will make it reconsider spending so much time working. The household selects a combination of leisure time and food to maximize its utility.

Each household has its own unique preferences regarding how much leisure and food it likes to consume. From any given leisure-food combination, the household would be willing to sacrifice some leisure time to work, but only if it is adequately compensated with additional food. There are an infinite number of work-food combinations that make the household equally well off, and these can be depicted by an indifference curve.

Indifference curves map the household's trade-off between consuming food and supplying time to work. Unlike a conventional indifference curve, which maps the trade-off between two desirable goods and slopes downward, an indifference curve that describes the preference trade-off between consumption and work slopes upward, because more work must be compensated with more food.

Fig. 5.2 presents three representative indifference curves for a household. Each indifference curve maps an infinite set of labor-food consumption pairs that make the household equally content. The household is said to be indifferent between any consumption-labor pair on the same curve. The household's utility, or satisfaction, is unchanged at any point along a given indifference curve, and its utility rises as the household skips to each progressive indifference curve, one above the other. These indifference curves are actually projections of slices of the utility function, $U(L_f, X_c)$, at given levels of utility onto the L_f, X_c plane, given that $l_c = \overline{T} - L_f$. Slices at higher levels of utility project

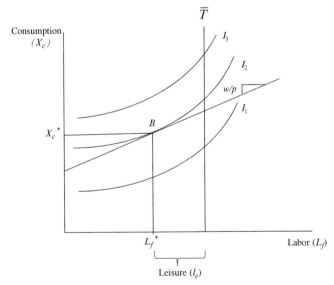

FIG. 5.2 Whereas a conventional indifference curve slopes downward, indifference curves describing the preference trade-off between consumption and work slope upward. The household's budget constraint slopes upward, as work generates cash for the household. The budget constraint's positive intercept reflects other income the household might have. This household's optimal consumption bundle is at point B, where its MRS equals the wage-to-price ratio.

indifference curves that lie farther to the northwest (more consumption, less work) on this plane. Thus, in Fig. 5.2, I_1 corresponds to a lower level of utility than I_2 or I_3.

The units of food required to compensate a household for employing an additional unit of labor so that the household is equally well off, given its own preferences for consumption and labor, is the *marginal rate of substitution* (*MRS*). The slope of the indifference curve is the consumer's *MRS*, which changes as one moves along the curve due to decreasing marginal utility from consuming more of any particular good. When, for example, the household works many hours, it requires a large quantity of food to compensate for working an additional unit. Conversely, when the household works few hours and it consumes relatively few units of food, then it is willing to work an additional hour of time in exchange for few units of food. The curvature of indifference curves depends on how quickly the household's marginal utility declines as it consumes less food or supplies more time to labor.

When there are efficient markets, the household maximizes its utility by consuming at the point where its *MRS* between food and work equals the ratio of the value of time (the wage) to the price of food, that is, w/p. At this point, the trade-off between working an additional unit of time and consuming food, given the household's preferences, just equals the additional amount of food the household could purchase at price p by working an additional unit of time at wage w. (As we shall see, when there are no markets, a shadow wage and a shadow price take the place of w and ρ, respectively.) In Fig. 5.2, the optimal allocation of labor and consumption, given the household's preferences, is illustrated by L_f^* and X_c^*, at point B on indifference curve I_2.

PUTTING PRODUCTION AND CONSUMPTION TOGETHER: AN AUTARKIC CHAYANOVIAN FARM HOUSEHOLD

Often markets in rural communities are incomplete or missing entirely. Sometimes the transaction costs of trade are so high that the household must consume what it produces without trade. In that case, there is a direct trade-off between enjoying more leisure and producing more food.

The father of the agricultural household model was a Russian agrarian economist named Alexander Chayanov. Chayanov believed that peasant households only produce enough food to survive. This position ultimately cost him his life. After the socialist revolution of 1917, Joseph Stalin wanted to set up large collective farms in Russia. Chayanov's theories suggested that these farms would not be efficient, because it would be hard to force small peasant farmers, accustomed to subsistence production, to cooperate and produce a surplus. Chayanov was arrested and shot on October 3, 1937.

A Chayanovian farm household is *autarkic*, because it produces only what it consumes and it employs labor only from its own time endowment—there is no hired labor. It exists in its own little closed economy. There are no market prices, because the autarkic household does not participate in markets. However, this doesn't mean that autarkic households do not put a value on food and labor—they do. Subjective valuations guide production and consumption decisions. We cannot see these prices, but we can see their influences or "shadows" on the decisions a Chayanovian household makes. Hence, the price of the food that a Chayanovian household produces is called the *shadow price*, and the wage is a *shadow wage*. You can think of the shadow price of food as being the price an autarkic household would be willing to pay to get a bit more food—if it could. And the shadow wage is the price it would be willing to pay to have a little more time.

Why might a labor market not exist for an agricultural household? One reason is the high cost of monitoring workers' effort in the fields (it is easier to monitor and motivate your own family members, who benefit directly from their own effort level, than a hired worker, who may have motives to shirk or "free ride" on other employees' work). Another is the high cost of workers traveling long distances to work on farms, particularly in poor countries with bad roads and no public transportation. Markets for agricultural output may be missing, too, if households are located far from a market, do not have access to transportation, and cannot communicate with potential buyers, thus making trade prohibitive.

Chayanov did not construct a formal agricultural household model, but other economists who followed him have.[2] Their models give insight into what labor supply and demand look like in a Chayanovian household.

In a simple economy with only one output, food (X), and one variable input, labor (L), a missing market for either food or labor forces the household to consume only what it produces. It produces where the $MRT = MRS$, that is, where the production function is tangent to the household's indifference curve. This is illustrated by point C in Fig. 5.3. The slope of the production function at this point is the ratio of the household's shadow price for leisure to its shadow price for food. Let ω (the Greek letter *omega*) be the shadow price for leisure and let ρ (*rho*) be the shadow price for food. Shadow prices are not observed, but their ratio reflects how much the household values an additional unit of leisure compared with an additional unit of food. When there is no trade, shadow prices and shadow wages may be different for all households in a village, depending on household endowments of time and capital as well as preferences.

In the Chayanovian model, consumption depends on production and vice versa, so household production and consumption cannot be determined separately. Prices in this model are endogenous to the system, like in any closed economy (e.g., a country without trade). In this case, production and consumption are said to be nonseparable and they must be derived simultaneously.

---

2. Chihiro Nakajima (1969) was a pioneer in the theory of agricultural household modeling.

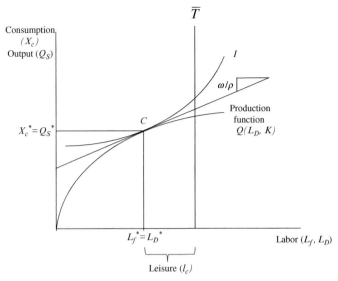

FIG. 5.3 In the Chayanovian model, a missing market for either the good or labor forces the household to consume only what it produces. The optimal production and consumption for this *autarkic* household are at point C, where the *MRT* = *MRS*.

LABOR IN AN AGRICULTURAL HOUSEHOLD MODEL WITH WELL-FUNCTIONING MARKETS

It is easy to illustrate that households can be made better off when they can trade with one another. In fact, households are never worse off by gaining access to trade. When markets function well, prices are determined exogenously by the aggregate supply and demand in the market, and all households in a market share the same prices and wages. Market prices and wages replace household-specific shadow prices and shadow wages. Since labor and food are traded in markets where wages and prices are given, household production and consumption decisions can be derived sequentially.[3] Income (and consequently the budget constraint on consumption) is determined by the output choices the household makes, given market prices. The prices of goods are unchanged regardless of the household's production and consumption decisions.

The household produces the optimal level of food, $Q_S{}^*$, using $L_D{}^*$ units of labor. The optimal production point is where the *MRT* is equal to the ratio of market prices. This point determines the household's income. From the production point, the household can afford any combination of food and work along the tangent line with slope w/p. The household purchases labor either from itself or

---

3. The basic references on agricultural household models with markets are: Barnum and Squire (1979) and Singh et al. (1986).

by hiring labor from the market at the market wage. Family and hired labor are perfect substitutes in this perfect-markets model.

The household can trade both food and labor at their market prices in order to consume an optimal (utility-maximizing) amount of food and leisure, given its budget. The household consumes X_c^*, while supplying L_f^* units of its own labor, which is where its *MRS* is equal to the ratio of prices.

Figs. 5.4 and 5.5 illustrate household production functions with two different potential consumption outcomes for different consumer preferences. Notice that, in these two figures, the household's optimal production of food and labor employed are unchanged regardless of the household's preferences. This is because the household can purchase labor and food in the market and thus decouple its production from its leisure and food demand. Notice also that household utility is higher than it would be if the household were forced to be self-sufficient, consuming at point A.

When the household provides its own labor for production, it essentially purchases labor from itself at the market wage. If the quantity of labor supplied by the household, L_f^*, is less than the household's demand for labor, L_D^*, then it hires in the difference. We refer to this additional labor not supplied by the household as hired labor, L_h^*. In the case shown in Fig. 5.4, the household is a net buyer of labor. The household with this indifference curve is a net seller of food, because it consumes less food than it produces, selling the surplus on the market.

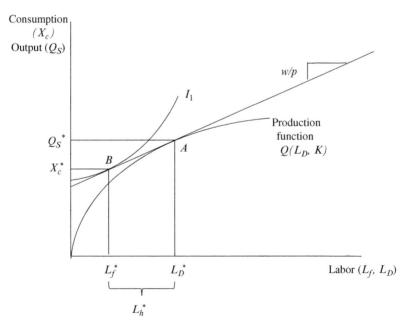

FIG. 5.4 Given market prices, this household produces at point A and demands an amount of labor equal to L_D^*. It consumes at point B and hires in an amount of labor equal to L_h^*. It is a net buyer of labor and seller of food.

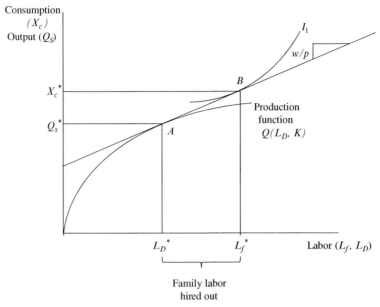

FIG. 5.5 Given market prices, this household produces at point A and demands an amount of labor equal to L_D^*. It consumes at point B and supplies more labor than it demands. It is a net seller of labor and buyer of food.

Conversely, if the household prefers to provide more labor units than it demands for its own production, it can hire out the additional units of labor to other producers in the market, as illustrated in Fig. 5.5. This household consumes more food than it produces, so it is a net buyer of food.

As mentioned previously, when the markets for food and labor are well functioning, households can buy and sell labor and food at fixed prices, and the agricultural household model is said to be separable, or recursive. This means that production decisions can be derived separately from consumption decisions. The household's consumption of food does not depend on its production, and vice versa. If the household demands more than it produces, it can purchase the difference, and if it produces more than it demands, it can sell the difference. Consequently, preferences for food or leisure have no impact on production decisions.

The household will always consume a combination of food and leisure such that the indifference curve's slope equals w/p. However, the indifference curve to which the household can aspire depends on the budget constraint. Higher profit from production shifts the budget line upward to the left, enabling the household to climb onto a higher indifference curve (to the northwest—less work, more consumption), which is associated with greater utility. The impact of higher wages on the household's budget and utility depends on whether the positive effect on potential wage income, wT, exceeds the negative effect on profit from food production.

THE EVOLUTION OF AGRICULTURAL HOUSEHOLDS

As economies develop, agricultural households become increasingly integrated with markets for food and for labor. We have seen that agricultural households operate more efficiently when they have access to markets, because they are not constrained to consume what they produce and they are not required to supply all of their own labor. Households that are highly endowed with time can trade their labor for more of the consumer good. Likewise, agricultural households with a small endowment of time can hire workers and sell their excess production in the market.

Market access is a critical component of economic and agricultural development, but it is important to keep in mind that market forces affect agricultural households in ways that are sometimes difficult to predict, since households both produce and consume the same goods.

Household Response to Price Changes

When households buy and sell goods, they are subject to the prices determined in the market. Integration with regional and global food markets can bring farmers higher prices for the food they produce. A striking example is when the completion of the transcontinental railroad in 1869 opened up nationwide markets for California's produce. This stimulated massive expansion of California agriculture. For a time, agriculture became the leading sector in the California state economy.

On the other hand, market integration could expose farmers to outside competition for the food they produce and lower prices. When California produce began arriving by the trainload to states in the Midwestern and eastern United States, some farmers there no doubt suffered as a result.

When the price of a good rises or falls, households adjust production and consumption accordingly. Let us use our model to consider how production, labor supply and demand, and consumption adjust to a rise in the price of food.

When the price of food rises, the household employs more labor to produce more food because the *MVPL* rises.

Fig. 5.6 illustrates a change in the price of food from p to p'. Initially, the household produces at point A and consumes at point B. In this figure, like Fig. 5.4, the household is a net seller of food and it hires additional labor from other households. When the price of food rises, the slope of the budget line becomes less steep. Production moves from point A to A'. The household hires more labor and produces more food. Consumption moves from B to B', where the *MRS* is equal to the slope of the new budget line. At B', the household consumes more food, but it works fewer hours because its demand for leisure, like other normal goods, goes up when its income rises.

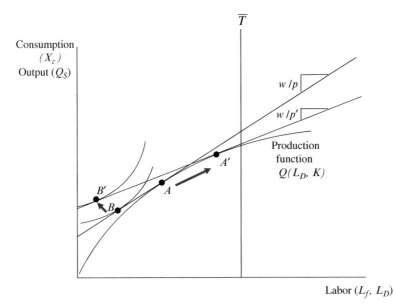

FIG. 5.6   An increase in the food price from p to p' induces the household to produce more food (at point A') while supplying less of its own labor (B').

Net sellers of food benefit from a rise in the price of food. This is not the case for net buyers. In most cases, if the household is initially a net buyer of food, it will shift to a lower indifference curve after the price increases, supplying more labor and consuming less food. Some households that initially have a small deficit of food may switch from being a net buyer to being a net seller when the price rises. The effects on utility for such households are ambiguous.

Impacts of Rising Wages on an Agricultural Household

The movement of labor off farms and into cities is a quintessential feature of economic development. When workers move out of agricultural areas, farm wages increase unless farmers can find a new source of labor (for instance, from immigration). The impacts of a rise in wages are difficult to analyze, because there is an impact on production (and demand for labor) as well as on leisure.

As a consumer, the opportunity cost of leisure rises with the wage, and the household substitutes food for leisure (i.e., it works more). This is the substitution effect. However, there is also a dual income effect, which pulls labor in opposite directions.

The household's potential income increases with higher wages because of the same endowment effect we learned about in Chapter 3. Its initial endowment of time increases in value, and as an owner of labor, the household's income rises when wages rise.

On the other hand, labor is an input into the household's production of the agricultural good. As a producer, the household loses when wages rise. Since the marginal cost of producing food increases with wages, the household reduces the amount of labor it employs. Lower production with higher production costs translates into less income.

Assuming leisure is a normal good, changes in leisure demand move in the same direction as changes in income. As wages rise, there are two income effects: one increases the demand for leisure because the household's time endowment increases in value, and the other reduces the demand for leisure because income from agricultural production decreases. Meanwhile, the substitution effect reduces the demand for leisure, because leisure is costlier relative to the consumer good. Consequently, a rise in rural wages can either increase or decrease the household's labor supply and have unintended welfare effects, depending on whether the household is a net seller or buyer of labor.

Welfare Implications

Since agricultural households both produce and consume the same goods, policies designed to improve household welfare through price changes often have ambiguous effects. In economies with well-functioning markets, welfare gains depend on whether the household is a net buyer or seller of the good in question. When households are outside the market, price changes have no effect on welfare outcomes. For price policies to affect household welfare, households must first gain access to well-functioning markets. This may require improvements in transportation to reduce marketing costs. It also may require improvements in communications (e.g., cell phones), so that households can get information about prices and about where and when goods can be bought and sold.

When there are multiple goods in the economy, in addition to a labor market, and only one market is missing, the household can still participate in trade. However, the impacts of a price change on household production and consumption are not straightforward. Consider an economy with perfect markets for two production goods and no market for labor. Households with a relatively high endowment of time allocate more resources toward producing labor-intensive crops. They sell labor-intensive crops and buy crops that require relatively less labor to produce. Conversely, households with a relatively small endowment of time allocate more resources toward producing less labor-intensive goods, which they sell in the market in exchange for more labor-intensive goods. That is, households focus on the goods for which they have a comparative advantage in production. This demonstrates the complexity of interactions between trade and price changes in an economy with a mixture of missing and perfect markets.

Migration, Education, and the Agricultural Transformation

Missing markets influence production and trade of goods. They also influence labor outcomes, including employment, education and migration. If a household does not have access to hired labor markets and its time endowment is relatively low, its shadow wage of labor may be very high. This may prevent household members from obtaining education—children are working in the fields instead of studying in the classroom. It also may prevent people from seeking off-farm work at a distant location. It can have important implications for welfare and development, since years of education and off-farm employment are both associated with higher earnings in rural communities (Duflo, 2000; Foster and Rosenzweig, 1996; Zhang et al., 2002; de Janvry and Sadoulet, 2001).

To see how labor markets impact education, consider an autarkic household with young children. The household must choose whether to send its children to school or employ them on the family farm. With many mouths to feed and few adult workers, the household's shadow price of food and its shadow wage may be high. The opportunity cost of leisure (or of going to school) might be not having enough food on the family's table. Parents may choose to pull children out of school if the shadow wage for children's labor on the farm is greater than the discounted future expected returns from going to school.

Shadow wages may not be constant year round. They may be higher during harvest. However, pulling children out of school for the harvest season may make it difficult for them to advance their education. The high opportunity cost of education for agricultural households without access to well-functioning markets may be a critical barrier to economic development, since education is a key to economic growth and access to off-farm employment (Nelson and Phelps, 1966; Mincer, 1984; Barro, 2001; Becker et al., 1994; Benhabib and Spiegel, 1994).

Migration is also an important determinant of household welfare. Migrant remittances to developing countries were $466 billion in 2017, according to The World Bank. Nobody knows how much money *internal* migrants send home. Migration can improve welfare in rural households through many potential avenues, including by providing a source of credit or insurance, spreading economic risk across geographies, or allocating household labor to locations and industries where it is more efficiently employed.

Many rural households have access to local labor markets that determine the wages they see but are isolated from outside, regional or national labor markets. The interaction of local (often, village) labor supply and demand then determines the wages people receive for doing local farm work or the wages they must pay the workers they hire. Isolated markets often lack the efficiency and potential for specialization that are possible with more open market access.

Improvements in transportation and communications infrastructure can link villages and households with outside labor markets. So can migrant networks, or contacts with relatives and friends who have migrated. Migrant networks can lower the costs and risks of migration by providing villagers with crucial information about jobs at migrant destinations. Frequently, people who have migrated offer new migrants assistance in finding employment and, in the case of unauthorized international migration, contacts with labor smugglers ("coyotes") to get across international borders. They may even provide new migrants with the funds they need to finance the border crossing, as well as insurance, by paying the coyote once the new migrant safely arrives at her destination. The migrant, in turn, may repay the loan once she begins working.

Farmers and farm labor contractors in rich countries rely heavily on "word of mouth" referrals through migrant networks to secure new workers from rural households abroad. Because of this, researchers refer to international migration as being a "network-driven process" (Massey et al., 1999).

The wage in migrant-sending communities (or shadow wage in households with missing markets) determines the reservation wage, or the net wage (after adjusting for migration costs) required to induce an individual to migrate. If a household is well endowed with labor but has a relatively small and unproductive farm, or if the wage in its village is low, there will be an incentive to send family members off as migrants—particularly if the household has access to migrant networks. Farmers at the migrant destination then might not have to offer a high wage in order to secure migrant labor.

In contrast, a high reservation wage discourages people from migrating. If rural Mexicans have access to nonfarm employment in their home country, farmers in Mexico and the United States will have to compete with nonfarm employers in both countries to secure agricultural workers, paying high farm wages.

CONCLUSIONS

Historically in today's high-income countries and today in the world's developing nations, most agricultural producers have also been households. Agricultural household models are hybrids of consumers and producers who hire and supply labor and consume all or part of the food they produce.

The role of labor in agricultural households—and the households themselves—evolve in the course of the agricultural transformation. We can imagine a continuum from autarkic to commercial agricultural households. Autarkic households supply all of their own labor and consume all of the food they produce. At the other extreme, most commercial agricultural households in the rich farm regions of California hire most of their labor and consume little if any of their output. As agriculture develops, agricultural households look more and more like firms and less like households.

Lewis (1954; see Chapter 3) illustrates how industrial growth raises the demand for food (and consequently the price of food) produced in the agricultural sector. Meanwhile, advanced technologies and improved capital in the agricultural sector make farm labor more productive, reducing the demand for agricultural workers. In the Lewis model, there is a simultaneous push and pull of labor off of the farm. A reduced demand for agricultural labor pushes workers off the farm. Industrial growth, rising wages, and increased demand for manufactured goods stimulate the demand for nonfarm workers, pulling labor out of agriculture. Growth in both the farm and nonfarm sectors is crucial for creating a labor force that is less agriculture-centered over time.

The expansion of regional, national, and international markets for labor and goods is an essential part of this transformation. Many high-income nations that have already undergone a far-reaching agricultural transformation employ farm workers from less-developed nations, where wages are relatively low. Migration networks connect household farms in developing countries with commercial farms in developed countries. However, as the migrant-source countries develop, the supply of farm workers can be expected to decrease there as well. Then, farmers at home and abroad may have to compete for a declining supply of farm workers.

APPENDIX

Mathematically Modeling Production and Consumption Decisions of Agricultural Households

Agricultural Households as Producers: Deriving Labor Demand

Baseline Profit Maximization: *MRT* = Ratio of Wage to Price

As producers, agricultural households choose how much labor to employ in order to maximize profits, similar to what we saw in Chapter 2. Profits are equal to the total revenue net the cost of inputs. In our model, total revenue is the per-unit price that can be obtained from selling the agricultural good times the quantity of the agricultural good produced, $pQ(L_D, \overline{K})$. Labor costs are equal to the market wages times labor demanded, wL_D. Households solve the maximization problem

$$\max_{L_D} \pi = pQ(L_D, \overline{K}) - wL_D.$$

To find the profit-maximizing quantity of labor demanded, we must derive the first-order conditions (FOCs). Take the derivative of profits with respect to labor and set it equal to zero

$$p\frac{dQ(L_D, \overline{K})}{dL_D} - w = 0$$

The FOC can be written as

$$p\frac{dQ(L_D, \overline{K})}{\partial L_D} = w.$$

The left-hand side of this expression is the *MVPL*, and the right-hand side is the market wage. That is, the household employs labor until the *MVPL*, $p\frac{dQ(L_D\overline{K})}{dL_D}$, is equal to the marginal cost of labor, w.

Dividing both sides of this expression by the price gives

$$\frac{dQ(L_D, \overline{K})}{dL_D} = \frac{w}{p}.$$

The left-hand side is now the *MRT*, or the additional quantity of the agricultural good that would be produced by employing one more unit of labor. This shows that the profit-maximizing household employs labor at the point where the *MRT* is equal to the ratio of wages to the price of the agricultural good.

From this expression, it is apparent that an increase in the price p causes the household to employ more labor and produce more of the agricultural good. This result follows from diminishing marginal returns to labor. An increase in p causes the right-hand side of the above equation to decrease. To retain equilibrium, the left-hand side of the equation must also decrease. By diminishing marginal returns to labor, $\frac{dQ(L_D,\overline{K})}{dL_D}$ decreases as L_D rises. Thus, to lower the *MRT*, more labor must be employed. Employing more labor increases total output, but it decreases the marginal product of labor. Taken together, this means that households employ more labor and increase production of the agricultural good when the price for the agricultural good goes up.

Deriving Labor Demand from a Cobb-Douglas Production Function

The profit-maximizing decisions of an agricultural household with access to markets can be demonstrated using a Cobb-Douglas function. Suppose that the household produces the agricultural good according to the function $Q = AL_D^{\alpha}\overline{K}^{1-\alpha}$, where Q is the quantity of the agricultural good produced, A is a shift parameter, L_D is the amount of labor employed, $\overline{K}$ is the fixed amount of capital used in production, and α is a parameter describing labor's share of production relative to capital. Let $0 < \alpha < 1$.

The Cobb-Douglas function is a useful example because it fulfills the fundamental assumptions of production theory. Production is increasing in labor and capital with diminishing marginal returns. To see these properties of the Cobb-Douglas function, first take the derivative with respect to labor

$$\frac{dQ(L_D,\overline{K})}{dL_D} = \alpha AL_D^{\alpha-1}\overline{K}^{1-\alpha} > 0$$

The first derivative of the Cobb-Douglas function with respect to labor shows that production is increasing in labor. Next, take the second derivative with respect to labor to understand the curvature of the production function

$$\frac{d^2Q(L_D,\overline{K})}{dL_D^2} = \alpha(\alpha-1)AL_D^{\alpha-2}\overline{K}^{1-\alpha} < 0$$

The negative second derivative of the Cobb-Douglas function with respect to labor shows that there are diminishing marginal returns to labor. We know that the above expression is less than zero because $\alpha < 1$.

Now, let us solve for the profit-maximizing quantity of labor employed

$$\max_{L_D} \pi = pQ(L_D,\overline{K}) - wL_D$$

Substituting in the production function

$$\max_{L_D} \pi = pAL_D{}^{\alpha}\overline{K}^{1-\alpha} - wL_D$$

The FOCs are then

$$\frac{d\pi}{dL_D} = p\alpha AL_D{}^{\alpha-1}\overline{K}^{1-\alpha} - w = 0,$$

which can be rewritten as

$$\frac{p\alpha}{L_D}AL_D{}^{\alpha}\overline{K}^{1-\alpha} - w = 0$$

Part of the above expression exactly resembles the production function for quantity of the agricultural product supplied $Q_S = AL_D{}^{\alpha}\overline{K}^{1-\alpha}$. We can simplify the expression by substituting in Q_S. (However, keep in mind when doing comparative statics that the quantity Q_S is endogenous to the amount of labor employed.)

$$\frac{p\alpha}{L_D}Q_S - w = 0$$

Rearranging the terms, we obtain the following expression for the profit-maximizing quantity of labor employed:

$$L_D{}^* = \frac{p\alpha Q_S}{w}$$

This expression shows that changes in the price of the agricultural good, p, impact the quantity of labor employed directly, since p enters directly into the expression, and indirectly, through p 's impact on the quantity of agricultural good produced, Q_S.

Agricultural Households as Consumers

Baseline Utility Maximization

Now, let us derive the demand for leisure (and by association, the family supply of labor) and the demand for food. Let l_c represent hours of leisure that the household consumes. It has a shadow value, or opportunity cost, equal to the market wage rate, w, since that is the forgone income that the household would have derived from working an additional hour. Each household is endowed with $\overline{T}$ units of time, which it allocates between labor and leisure. Let X_c represents the quantity of the agricultural good that the household consumes. Let the household's full income be denoted by Y, where Y is the household's profit from producing the agricultural good plus the market value of the household's time endowment. This determines the household's budget constraint. The household cannot spend more resources on the agricultural good or leisure than its full income, Y, affords. Let $Y = w\overline{T} + pQ_S - wL_D$, where Q_S is the quantity of the agricultural good produced.

The household's utility function is $u(l_c, X_c)$, where $\frac{\partial u}{\partial l_c} > 0$ and $\frac{\partial u}{\partial X_c} > 0$. That is, utility is increasing in leisure and consumption of the agricultural good, X_c. Further, let $\frac{\partial^2 u}{\partial l_c^2} < 0$ and $\frac{\partial^2 u}{\partial X_c^2} < 0$, and let $\lim_{l_c \to 0} \frac{\partial u}{\partial l_c} = \infty$ and $\lim_{X_c \to 0} \frac{\partial u}{\partial X_c} = \infty$. This simply means that there are diminishing marginal returns to consumption of either good, and the marginal utility for either good rises rapidly as the consumption of that good is driven to zero (households must consume at least a little bit of each good, both leisure and food). These assumptions assure an interior solution. (To derive Figs. 5.2–5.6, we substituted for l_c from the time constraint to express utility as a function of time worked instead of leisure.)

The household maximizes utility with respect to leisure and the consumption good, subject to its budget constraint. The household's utility-maximizing problem can be expressed.

$\max_{l_c, X_c} u(l_c, X_c)$, subject to $w l_c + p X_c \leq w \overline{T} + p Q_S - w L_D$.

Since we know that there is an interior solution, we can express the Lagrangian as

$$\max_{l_c, X_c} u(l_c, X_c) - \lambda \left(w l_c + p X_c - w \overline{T} + p Q_S - w L_D \right)$$

and the FOCs are

$$\frac{\partial u(l_c, X_c)}{\partial l_c} - \lambda w = 0,$$

$$\frac{\partial u(l_c, X_c)}{\partial X_c} - \lambda p = 0,$$

and

$$w l_c + p X_c = w \overline{T} + p Q_S - w L_D.$$

Combining the FOCs describing the marginal utilities of leisure and food results in the expression

$$\frac{\dfrac{\partial u(l_c, X_c)}{\partial l_c}}{\dfrac{\partial u(l_c, X_c)}{\partial X_c}} = \frac{w}{p}.$$

The left-hand side of the above equation is the household's *MRS*, or the amount of leisure it would be willing to give up in exchange for an additional unit of the agricultural good. At the optimum, the household consumes leisure and food at the point where the *MRS* is equal to the ratio of wage to price.

We have just derived an expression to quantify the agricultural household's trade-off in preferences between leisure and consumption of the agricultural good. Where the *MRS* is equal to the ratio of the wage to the price for food, the trade-off in preferences is equal to the market trade-off. There are many indifference curves that the household can choose from. The household maximizes its utility by spending all of its budget. This is the full-income constraint, which is the final FOC: $wl_c + pX_c = w\overline{T} + pQ_S - wL_D$. This budget constraint is contingent upon the profit-maximizing quantity of agricultural good produced, Q_S.

Deriving Consumption Demand From a Cobb-Douglas Utility Function

If the household's preferences can be represented by a Cobb-Douglas utility function, $U(l_c, X_c) = l_c^{\beta_1} X_c^{\beta_2}$, then the FOC for utility maximization simplifies to

$$\frac{\beta_1 X_c}{\beta_2 l_c} = \frac{w}{p}$$

subject *to* the budget constraint $wl_c + pX_c = w\overline{T} + pQ_S - wL_D$

Let $\beta_1 + \beta_2 = 1$, a common assumption that implies that increasing leisure and food by a common factor increases utility by the same factor (i.e., doubling leisure and food consumption doubles utility). Rearranging, we can derive the consumer demands as linear functions of full income and inverse functions of the good's own price. Full income includes the profit from crop production, assuming optimality on the production side.

$$l_c = \frac{\beta_1}{w}\left(w\overline{T} + pQ_S - wL_D\right)$$

$$X_c = \frac{\beta_2}{p}\left(w\overline{T} + pQ_S - wL_D\right)$$

What are β_1 and β_2? Rearranging the consumption demand equations we can see that these are the budget shares of the two goods:

$$\beta_1 = \frac{wl_c}{w\overline{T} + pQ_S - wL_D}$$

$$\beta_2 = \frac{pX_c}{w\overline{T} + pQ_S - wL_D}$$

Assuming Cobb-Douglas utility is convenient, because the budget shares are easy to calculate from household data, and the consumer demands are linear in income. Is Cobb-Douglas utility a reasonable assumption? What does it imply about the household-farm's preferences?

Cobb-Douglas utility functions have some theoretical advantages as well as being easy to calculate. Utility is increasing in consumption with diminishing marginal returns, and multiple consumption goods are complementary. That is, if an individual has a large share of good 1 and not very much of good 2, then the individual would be willing to part with quite a bit of good 1 to acquire a little more of good 2. Likewise, the individual would require a large compensation of good 1 if you asked him or her to part with a little of good 2. All of these attributes can easily be captured in a Cobb-Douglas utility function, making it an attractive choice for economic models.

The Nonseparable Optimization Problem for the Autarkic Household

When households do not have access to markets, then they cannot buy or sell excess labor or food in a market, and they do not observe market prices w or p. They must provide all of their own labor to agricultural production, and they must consume exactly what they produce. In the autarkic household, production and consumption are nonseparable.

Let household production be characterized by the function $Q_S = Q(L_D, \overline{K})$, where $\frac{dQ(L_D, \overline{K})}{dL_D} > 0$ and $\frac{d^2Q(L_D, \overline{K})}{dL_D^2} < 0$. That is, production is increasing in labor with diminishing marginal returns, the same assumptions that we made earlier.

Let the household utility be defined by $u(l_c, X_c)$, where $\frac{\partial u}{\partial l_c} > 0$, $\frac{\partial u}{\partial X_c} > 0$, $\frac{\partial^2 u}{\partial l_c^2} < 0$, and $\frac{\partial^2 u}{\partial X_c^2} < 0$. Further assume that $\lim_{l_c \to 0} \frac{\partial u}{\partial l_c} = \infty$ and $\lim_{X_c \to 0} \frac{\partial u}{\partial X_c} = \infty$. This assures an interior solution.

The household solves the optimization problem

$$\max_{l_c, X_c} u(l_c, X_c)$$

subject to $l_c \leq \overline{T} - L_D$ and $X_c \leq Q_S = Q(L_D, \overline{K})$. Notice that labor demand is now in the time constraint (because it equals the household's labor supply), and the household can consume only the food it produces.

The constraints will be binding since utility is monotonically increasing in leisure and consumption of food. We can therefore substitute the constraints into the optimization problem

$$\max_{L_D} u\left(\overline{T} - L_D, Q(L_D, \overline{K})\right)$$

The FOC is

$$\frac{\partial u}{\partial l_c}(-1) + \frac{\partial u}{\partial X_c}\frac{dQ}{dL_D} = 0,$$

which can be rearranged to

$$\frac{dQ}{dL_D} = \frac{\dfrac{\partial u}{\partial l_c}}{\dfrac{\partial u}{\partial X_c}}.$$

The left-hand side of the above expression is the *MRT* and the right-hand side is the *MRS*. Consequently, the autarkic household optimizes consumption and production by producing at the point where *MRT* = *MRS*.

Let ω be the implicit shadow wage, and let ρ be the implicit shadow price of food. The shadow prices ω and ρ are not observed directly. Rather, they are implied by the household's production and consumption decisions. If we could map the household's preferences into a utility function then we could derive the shadow prices. Instead, we only observe their ratio according to the relation

$$\frac{dQ}{dL_D} = \frac{\omega}{\rho}.$$

Production and Consumption for Households With Perfect Market Access

When households have access to markets, they can buy and sell goods at their market wages, so their valuations of labor and food are the market wage and price. Since the household's preferences do not determine the valuations for leisure (labor) and food, producer and consumer optimization are separable.

Producer theory shows us that households, performing as profit-maximizing firms, produce where the *MRT* is equal to the ratio of the wage to the price of the agricultural good. Consumer theory shows us that the household, performing as a consumer, maximizes utility by consuming at the point where the *MRS* is equal to the ratio of wage to price. Thus, *MRT* = *MRS* in equilibrium. However, since households have access to trade, they are not constrained to consume the same leisure-agricultural good pair that they produce.

Production does not depend on consumer preferences, but consumption depends on production to determine its budget constraint. The slope at the tangency point, where the consumer's indifference curve meets the budget constraint, is unaltered by production. It is determined only by preferences and market prices. Consequently, when there are perfect markets, we can solve the producer problem in the "Agricultural Households as Producers: Deriving Labor Demand" section of the Appendix, and then we can solve the consumer problem in "Agricultural Households as Consumers" section once we have determined the household's budget from its profit as a producer and owner of labor. This gives us, first, the household's production pair of labor demanded and agricultural good produced or supplied, and second, the household's consumption pair of leisure consumed (thus family labor supplied) and the agricultural good consumed.

By combining the results from the producer and consumer problems, we learn whether the household is a net seller or buyer of labor and of the agricultural good. If the household demands more labor to maximize profits than it optimally supplies as a consumer, it will hire additional labor in the market at wage *w*, making the household a net buyer of labor.

The amount of labor supplied by the household is family labor supply, denoted by $L_f = \overline{T} - l_c$.

The amount of labor that the household hires from the market is hired labor, denoted by $L_h = L_D - L_f$.

In the 2-good model, where households consume only leisure and a single agricultural good, households that are net buyers of labor produce more of the agricultural good than they consume. Thus, the household is a net buyer of labor and a net seller of food. The amount of food that the household sells in the market is the marketed surplus of food, denoted by $X_{surplus} = Q_S - X_C$.

Conversely, households that demand less labor than they supply will hire out their family labor in the market. These households are net sellers of labor. In the 2-good model, these households will demand more food than they consume, so they are net buyers of X, paying market prices.

REFERENCES

Barnum, H.N., Squire, L., 1979. An econometric application of the theory of the farm-household. J. Dev. Econ. 6 (1), 79–102.

Barro, R.J., 2001. Human capital and economic growth. Am. Econ. Rev. 91 (2), 12–17.

Becker, G.S., Murphy, K.M., Tamura, R., 1994. Human Capital, Fertility, and Economic Growth. In: Human Capital: A Theoretical and Empirical Analysis with Special Reference to Education. third ed. The University of Chicago Press, Chicago, pp. 323–350.

Benhabib, J., Spiegel, M.M., 1994. The role of human capital in economic development evidence from aggregate cross-country data. J. Monet. Econ. 34 (2), 143–173.

de Janvry, A., Sadoulet, E., 2001. Income strategies among rural households in Mexico: the role of off-farm activities. World Dev. 29 (3), 467–480.

Duflo, E., 2000. Schooling and Labor Market Consequences of School Construction in Indonesia: Evidence From an Unusual Policy Experiment. National Bureau of Economic Research, Inc. NBER Working Papers 7860.

Foster, A.D., Rosenzweig, M.R., 1996. Technical change and human capital returns and investments: evidence from the green revolution. Am. Econ. Rev. 931–953.

Lewis, W.A., 1954. Economic development with unlimited supplies of labor. Manchester School of Economic and Social Studies 22, 139–191.

Lowder, S.K., Skoet, J., Raney, T., 2016. The number, size, and distribution of farms, smallholder farms, and family farms worldwide. World Dev. 87, 16–29.

Massey, D.S., Arango, J., Hugo, G., Kouaouci, A., Pellegrino, A., Taylor, J.E., 1999. Worlds in Motion: Understanding International Migration at the End of the Millennium. Clarendon Press, Oxford.

Mincer, J., 1984. Human capital and economic growth. Econ. Educ. Rev. 3 (3), 195–205.

Nakajima, C., 1969. Subsistence and commercial family farms: some theoretical models of subjective equilibrium. Subsistence Agriculture and Economic Development 165, 185.

Nelson, R.R., Phelps, E.S., 1966. Investment in humans, technological diffusion, and economic growth. Am. Econ. Rev. 69–75.

Singh, I., Squire, L., Strauss, J., 1986. Agricultural Household Models: Extensions, Applications, and Policy. Johns Hopkins University Press, Baltimore.

Taylor, J.E., Lybbert, T.J., 2015. Essentials of Development Economics. University of California Press, Berkeley, CA.

Zhang, L., Huang, J., Rozelle, S., 2002. Employment, emerging labor markets, and the role of education in rural China. China Econ. Rev. 13 (2), 313–328.

Chapter 6

Farm Labor and Immigration Policy

The farmworkers of tomorrow are growing up today outside the United States.

Philip L. Martin

From Europe to New Zealand and even parts of Africa and Latin America, as domestic workers move out of agriculture, farmers and the labor contractors that serve them seek workers from lower-income countries abroad. In the United States, wave upon wave of immigrants fueled the expansion of labor-intensive crops, particularly in California and other western states. This chapter explains how immigration expands to fill an excess demand for farm labor. Farmers grapple with a political process that often is resistant to permitting large numbers of low-skilled foreign immigrants into their countries legally. Because of this, immigrant farm workers often are unauthorized, eking out a livelihood in places that demand their labor but where the threat of apprehension and deportation is a daily reality. Policies to reform this system frequently are a patchwork of compromises that produce unintended consequences for immigrant workers as well as for the societies in which they work.

The United States began its history as a nation of small agricultural households, but it has evolved to have the most dynamic commercial agricultural industry in the world. The vast majority of crop production today is on large capital-intensive grain or labor-intensive fruit and vegetable farms employing a foreign hired workforce, not small farms tilling the land with family labor. Few people born in the United States grow up to be farmworkers anymore, but American farms had the good fortune of having Mexico next door, where millions of people were willing to migrate northward to do hired farm work. In 2014, around three-quarters of all hired farmworkers in the United States were foreign-born, and in California, fewer than 10% of farmworkers were US-born (U.S. Dept. Labor, 2017). Other high income countries are not much different from the United States in this regard. It is rare to find a native Englishman working for a wage on a UK farm, a German on a German farm, or a Spaniard on a Spanish farm. To paraphrase Philip Martin, an expert on farm labor, the farm workforce of tomorrow is growing up on poorer, foreign soils.

The Farm Labor Problem. https://doi.org/10.1016/B978-0-12-816409-9.00006-9

Like in W. Arthur Lewis' model of the agricultural transition, most of the workforce in today's high-income countries shifted to urban jobs in the 20th century, leaving few willing to do farm work. The Lewis model predicts that the movement of labor off the farm bids up wages in agriculture, as farms compete with the urban sector for a limited supply of workers. Rising farm wages, in turn, create incentives for farms to adopt labor-saving technologies, just as other industries have done. There has been considerable technological change in agriculture around the world, but mostly it has had the goal of raising yields per acre of land, or to dramatically reduce labor demands in commercial staple crops like maize, wheat, and rice. Mechanization of cultivation and harvest processes is more challenging for delicate crops like fresh fruits and vegetables. Consequently, development of labor-saving technologies for these crops has lagged behind.

The incentives to develop and adopt labor-saving techniques have been limited by the availability of foreign agricultural workers willing to migrate and work on farms in relatively high-income countries. Instead of mechanization, producers of labor-intensive crops have turned to immigration, importing foreign-born farm workers to labor for wages that are low relative to nonfarm wages in high-income countries but high compared with farm wages in the immigrants' countries of origin. Farmers' demand for foreign workers frequently clashes with an anti-immigrant stance by other factions in society.

In no state was the historical conflict between farmers' pro-immigration stance and society's backlashes against immigrants more apparent than in California. Labor-intensive fruit, vegetable, and horticultural (FVH) production took root and expanded in California largely because of the availability of Chinese immigrant workers after the completion of the transcontinental railroad. These workers were unable to find work in Sacramento, San Francisco, and other cities because of discriminatory anti-immigrant labor laws. Unemployed Chinese were willing to work for low wages and did not demand housing or compensation between growing seasons, which made them less expensive to retain as workers from year to year.

Eventually, the Chinese were banned by a federal law and replaced by Japanese immigrants, who were replaced by Pakistanis and Indians, who were replaced by Filipinos, and then Mexicans. Mexican farm workers were temporarily replaced by "Dust Bowl" American migrant workers from Oklahoma, Arkansas, and Texas, who moved out west in the 1930s after their lives were devastated by a dramatic climate shock. Today, rural Mexico is the primary source of hired farm labor to the United States, even though the majority of Mexican farm workers are not authorized to work in the country legally. Farm labor advocates grapple with the question of how to ensure fair working and living conditions for a large, vulnerable population, mostly ignored by government.

This chapter discusses the role of immigration in the agricultural sectors of high income countries. As we saw in Chapter 4, even middle-income countries import farmworkers to toil in their fields, but immigrants are a pervasive feature of the farm workforce in all high-income countries of the world today. No country in the world offers the same detailed data on immigrants or as rich of a farm labor history as the United States, where wave upon wave of immigrant workers fed the expansion of high-value crops. While keeping a global perspective, this chapter focuses on the history of US agriculture, how the workforce evolved, and some of the potential consequences of immigration on wages, crop mixes, and agricultural expansion. Many of the insights we glean from the history of US farm labor are applicable to other high-income countries around the world.

IMMIGRATION AND FARM LABOR MARKET EQUILIBRIUM

In countries with a large urban sector, access to an elastic supply of foreign-born farmworkers can make the domestic farm workforce negligible in equilibrium. Consider Fig. 6.1, which is a good representation of the farm labor market equilibrium in the United States and other highly industrialized countries. The domestic farm labor supply is represented by the curve LS_{dom}' in panel (A). Its steep upward slope indicates that the domestic farm labor supply is inelastic. Few domestic workers are willing to work in agriculture. More

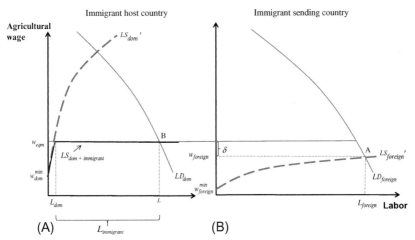

FIG. 6.1 Connected farm labor markets across countries. An elastic immigrant farm workforce can make the domestic farm labor supply inconsequential, keeping wages relatively low when the domestic farm labor supply is more inelastic. (A) The farm labor supply and demand in the domestic (host) country. Domestic farm labor supply is inelastic (steep), but employers can hire many immigrant workers and a few domestic workers if they pay only the foreign country's reservation wage. The equilibrium total workers employed and wage in the domestic farm labor market is at point B. (B) The farm labor demand and supply in the foreign (immigrant-sending) country. Farm labor supply is elastic (relatively flat), and the equilibrium wage at equilibrium point A is $w_{foreign}$. The cost of migrating to the immigrant host country is δ, and $w_{foreign}+\delta$ is the reservation wage, the minimum wage that workers are willing to accept to migrate. *(Modified from Taylor, J.E., Charlton, D., Yunez-Naude, A., 2012. The end of farm labor abundance. Appl. Econ. Perspect. Policy. 34(4), 587–598.)*

domestic workers are willing to work in agriculture as wages rise, but wages have to rise by a large amount to induce a marginal number of domestic workers to join the farm workforce.

Farm employers do not pay the high rates required to attract more than a few domestic workers into the fields. Instead, they employ a large share of immigrants who provide a more elastic supply of labor to farms at a relatively low reservation wage.

In Fig. 6.1, the flat, perfectly elastic line that intersects the domestic labor demand curve at the equilibrium wage w_{eqm} represents the immigrant farm labor supply. The immigrant farm labor supply is determined by the farm labor supply in the immigrant-sending (poorer) country, illustrated in panel (B). In this example, the labor supply in the sending country is perfectly elastic, like in a Lewis model. Farmers can hire as many workers as they wish as long as they pay a wage at least as great as the immigrants' reservation wage, the minimum wage required to attract foreign workers to migrate from their home country to work on rich-country farms. At the equilibrium shown by point A, $L_{foreign}$ workers work in the farm sector in the sending country at wage $w_{foreign}$. Workers from the poor country will migrate to the rich country's agricultural sector if wages there are at least as high as wages in the poor country plus the cost of migration, δ. The expression $w_{foreign} + \delta$ is the reservation wage. Equilibrium in the rich country's farm labor market is at point B in Fig. 6.1.

Unlike the agricultural household model in which households face high transaction costs and imperfect labor markets, the equilibrium in Fig. 6.1 reflects an environment with well-functioning markets and an abundant supply of immigrant labor that easily substitutes for household or domestic labor in the immigrant-receiving country.

IMMIGRANT WORKERS IN US AGRICULTURE[1]

In the 19th century, family farms were the ideal reality for most of US agriculture. The striking exception was the plantation system in the South, which relied on slave labor forcibly imported from Africa prior to the Civil War. Economic historians Robert Fogel and Stanley Engerman (1980) studied the economic rationale for slavery in the US antebellum south. The evidence they present strongly suggests that slavery was not only profitable but also more efficient than the employment of free labor. The Southern economy was growing at a healthy pace prior to the Civil War, labor demand was increasing, and this suggested that there was not an economic impetus to bring about an end to slavery. It was profitable for plantation owners to keep slaves year-round in large part because cotton and tobacco had long, laborious growing seasons, 6–8 months long. The price of slaves fluctuated with the price of tobacco, indicating that the institution of slavery was closely tied to its profitability.

---

1. This section draws heavily from Taylor (2010), Martin et al. (2006), and Martin (2003).

Plantations also relied heavily on imported slave labor in British and French colonies of the Caribbean to produce sugar for an expanding European market, but the European colonial powers abolished slavery before the United States did. After the United States ended slavery, the immigration of people with low reservation wages allowed US farms to continue operating and expanding their production through the employment of foreign labor at relatively low cost. This increased the demand for farm land, raising land rents. Thus, many of the benefits of cheap labor became capitalized in high land prices.

In 1869, completion of the transcontinental railroad opened up markets in the eastern United States for FVH produce from the western states, especially California. Labor demands on FVH farms differed from those on plantations in the South because labor demand was highly seasonal. Farms in the West relied on wage workers migrating from farm to farm to meet their seasonal labor demands. Wave upon wave of immigrants from a diversity of countries helped transform California into the richest agricultural region in the world, and these workers eventually fanned out across the entire country. US farm-labor history is the story of waves of newcomers entering the country to do farm work, and then returning to their country of origin or moving into nonfarm jobs. Farmworkers' children educated in the United States generally refused to follow their parents into the fields, so most new farmworkers were raised outside the United States (Martin, 1996). The one exception was a brief period in which California's farm workforce became mostly domestic again, but it took a cataclysmic climate shock, the Dustbowl, to make that happen.

FROM RAILROAD TO FARM WORKER

Completion of the transcontinental railroad in 1869 opened up new opportunities for fruit and vegetable production by giving Western farmers better access to consumer markets in the eastern US. Prior to the completion of the railroad, California farms had few markets for perishable crops. Fresh fruits and vegetables could be sold at higher prices than wheat, but they had to be sold shortly after harvest. It took too long to ship fresh fruits to Eastern markets by boat, but now perishable crops could be shipped by rail. Access to markets for perishable crops was further facilitated by the development of the refrigerated boxcar in 1888, one of which is on display today at the California State Railroad Museum in downtown Sacramento.

The railroad permitted Western farmers to ship their products more quickly and cheaply, but the refrigerator cars had to be filled. California's great Central Valley, which stretches 450 miles from Redding in the north past Bakersfield in the south, already had two of the four prerequisites to become one of the most productive agricultural regions in the world: It had fertile alluvial soils and plenty of sunlight. However, the production of fruits, vegetables, and horticultural crops is intensive in two other inputs that were in short supply: water and labor.

Californian farmers began investing in irrigation systems (with taxpayer support). Nature gave California the largest natural reservoir imaginable. Historically, in a normal year, the snowpack in the Sierra Nevada Mountains stores 15 million acre-feet of water well into the dry summer months (Chou, 2014). (An acre-foot is about 360,000 gallons of water, enough to cover a football field one foot deep. In 2011, a California household consumed an average of a little over a third (0.36) of an acre-foot of water annually (Aquacraft Water Engineering and Management, 2011.)) As the snow gradually melts, it fills river basin systems just as crops are beginning to be planted.

Fruit and vegetable production required not only abundant water but also a large seasonal workforce to manage and harvest crops by hand. If you were a California FVH farmer, the ideal would be to flood your fields with cheap water in the growing season and cheap labor in the harvest season. Thanks to irrigation, farmers had access to enough cheap water to make deserts bloom. With few domestic workers willing to do farm work in the sparsely populated West, though, access to cheap labor required immigration.

Fortunately for large land holders, the railroad provided access not only to Eastern markets but also to a large, low-wage immigrant workforce. More than 12,000 Chinese were brought into the United States to work on the transcontinental railroad, and after the railroad was completed almost all of them were in search of new employment (Martin, 2003). Many Chinese immigrants tried mining, but American claim jumpers usually drove them away. Discriminatory laws made it difficult, or even impossible, for the Chinese to find employment in the cities. Having no other option, Chinese immigrants turned to the agricultural sector. They were willing to work at lower wages than US-born workers, and they were willing to work seasonally. This seasonal labor supply, available at low wages, was the final piece that landowners needed to make fruit and vegetable production explode in California. By the early 1880s, Chinese workers constituted 75% of California's seasonal farm labor force (Fuller 1939/40).

Fig. 6.2 illustrates what the California agricultural labor market might have looked like at the completion of the transcontinental railroad. Initially, California grew wheat and other grain crops that required few workers. When farmers gained access to markets in the East, they began planting orchards and other perishable crops, and the demand for farm workers shifted outward from LD to LD'. Concurrently, the completion of the railroad left thousands of Chinese workers unemployed. Consequently, the total labor supply curve rotated outward from a relatively inelastic supply of domestic workers, illustrated by curve LS_{Total}, to a very elastic supply of Chinese workers, illustrated by curve LS_{Total}'.

Chinese were highly regarded for their self-sufficiency and willingness to work hard for little pay. One cotton farmer coming to California from the East reported that labor was abundant in California: "White men can be hired for one dollar per day, with board. Chinamen [sic] in any quantity at $25 per month, they boarding themselves" (Arax and Wartzman, 2005, 36). Chinese would work for lower wages than American field workers, and they would disappear in groups at the end of the work day, managing their own housing. They would

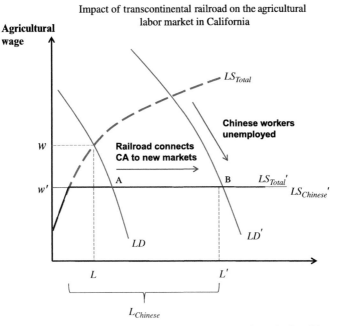

FIG. 6.2 This figure illustrates how the completion of the transcontinental railroad increased the demand as well as the supply of farm labor. Equilibrium farm employment increases from L *(at point A)* to L' *(point B)*.

further disappear after harvest until the beginning of the next growing season. For the landowner, this situation was ideal. Many argued that farmers could not afford to employ workers year-round:

> *California farmers cannot hope to employ Americans or European immigrants because farmers cannot employ them profitably...more than three or four months in the year—a condition of things entirely unsuited to the demands of the European laborer.*

> (Pacific Rural Press, quoted in Martin, 2003, 38).

Fresh fruit and vegetable production in California flourished on a foundation of inexpensive, seasonal labor, and the industry remained dependent on workers being available when needed at low cost.

Chinese workers had myriad impacts on California agriculture. They contributed not only physical labor but also knowledge and skill. Landowners in California knew little about orchard care prior to 1870, and Chinese workers brought experience in orchard management from the villages in China where they grew up. Sources indicate that Chinese workers taught landowners about orchard cultivation, suggestive that these new agricultural endeavors might not have succeeded without a knowledgeable workforce skilled in orchard care (Martin, 2003).

Despite all that Chinese immigrants contributed to California's advancing economy, many residents, particularly in the cities, were opposed to continued Chinese immigration. To California farmers' dismay, opposition to Chinese immigration mounted, culminating in the US Congress' passage of the Chinese Exclusion Act of 1882. It prohibited all immigration of Chinese workers. Chinese immigration came to a halt when President Chester Arthur signed the Act into law.

With few immediate alternatives, landowners continued to employ Chinese workers in opposition to the law. The 1892 Geary Act continued the rulings of the 1882 Exclusion Act and further curtailed the rights of Chinese immigrants. Some rural residents took the law into their own hands, driving Chinese workers from the fields at night, and it became difficult for Chinese immigrants to remain in California farm work.

JAPANESE WORKERS REPLACE THE CHINESE

The absence of an abundant, inexpensive workforce after the exclusion of Chinese workers threatened the viability of large-scale farming in California. The presence of low-cost immigrant workers had kept agricultural profits high and raised the value of agricultural lands. Around the same time that Chinese workers were excluded from the fields, fruit prices began to sag, and there was a brief trend toward smaller family farms in California. However, strategic planning by a monopoly of sugar beet processors and acquisition of a steady supply of workers from Japan quickly shifted agricultural production back to large farms.

Around the time of the California gold rush, an enterprising German immigrant named Claus Spreckels built a sugar beet factory in Watsonville. At the time, there was insufficient demand for sugar beets and the factory was soon shut and abandoned. However, the Dingley Tariff Act of 1897 put steep tariffs on imported sugar, and sugar beet production suddenly became more profitable. Shortly thereafter, sugar beet factories began to open around the state. Field laborers were not abundant at the time, and initially, sugar beet factories contracted with small family farmers to grow sugar beets and sell them to the factories for processing. However, the factories had a monopoly on the purchase of sugar beets, giving them leverage to eventually buy land from small farmers.

Large farms required a large workforce, and former small-farm owners were not as accommodating to the seasonal demands of large farms as Chinese workers had been. In 1885, Japan legalized emigration, and agriculture was a major sector of employment for new Japanese immigrants. Japanese workers dominated the farm worker population in California from 1900 to 1913.

Japanese farm workers had a few defining characteristics that distinguished them from other immigrant farm workers that came before or after. At first they worked for lower wages than both the white and Chinese farmworkers and disappeared as soon as the work was done, just as landowners desired. However, some had a longer-term agenda. Many managed to buy land, usually marginal lands like swamps that had to be cleared. This permitted Japanese workers to begin farming for themselves, which frustrated both large and small farmers. For large farmers this meant the rapid disappearance of what had been an inexpensive, elastic workforce. For small farmers it meant competition. Japanese buyers bid up land values of small farm parcels; they often were able to outbid competing native small farmers (McWilliams, 1939). Many Japanese workers, instead of earning wages, would bargain with farmers for a share of the crop. They would then sell the produce, potentially netting more than the wages they would have received and honing trade skills that would prove valuable once they began their own farms. Japanese produce shops became commonplace.

Eventually, in 1907, the United States made an agreement with the Japanese government to end Japanese immigration. Further restrictions were placed on Japanese immigrants through California's Alien Land Act of 1913, which prohibited "aliens ineligible for citizenship" from owning or having long-term leases on agricultural land, and the US Immigration Act of 1924, which banned virtually all immigration from Asia and increased enforcement. However, before immigration came to an end, the Japanese made a substantial impact on California's agriculture and overall economy. Japanese immigration allowed California's agricultural production area to increase from 1.4 to 2.7 million acres between 1899 and 1909. During the same period, the value of fruit production doubled (Martin, 2003). However, Japanese workers made more demands than the Chinese did. They were known for organizing strikes just before harvest, called "Quickie strikes." These strikes were effective because Japanese replacement crews refused to break the strikes of other Japanese workers. Strikes may not have brought about any large reforms, but they did raise Japanese workers' wages, and they posed a threat to the prevailing hierarchy of employment in California agriculture.

Following the Japanese, immigrants from Pakistan and current-day India came to work on Californian farms. They mostly arrived between 1907 and 1910. However, few remained in hired farm work, preferring to farm their own land, instead. Pakistanis and Indians were followed by immigrants from Armenia, and then from the Philippines.

Between 1917 and 1921, the United States entered into a guest worker agreement with Mexico that permitted Mexicans to migrate legally to work on US farms during the growing season and afterwards return to Mexico. It was the first in a series of agreements that became known as the Bracero Program. However, discriminatory practices against the Mexican Bracero workers were commonplace, and workers were able to save little by the time they returned home. The Mexican government expressed disappointment over how the agreement had turned out and showed no interest in continuing the Program.

During this time, there were some efforts in California to import domestic workers from the Eastern United States, and numerous transient workers (or hoboes) roamed around the state, following the crops and rarely saving enough money to buy their own piece of land. There were a few attempts to organize farm labor unions among these transient workers, but they were mostly unsuccessful, largely because it is difficult to organize workers who are always on the move and too poor to pay union dues.

DUSTBOWL MIGRATION CROWDS OUT IMMIGRANT LABOR IN THE FIELDS

In 1935, severe dust storms, fed by drought, ravaged the ecology and agriculture of the American prairie states, affecting 100 million acres in Texas, Oklahoma, Kansas, New Mexico, and Colorado (Hakim, 1995).[2] The only viable option for many families laid destitute by the drought, known as the Dustbowl, was to migrate. Dustbowl migrants poured into California by the thousands, living in tent camps called Hoovervilles. With few skills other than the ones their agricultural roots had given them, they moved from farm to farm following the crops. Their plight was immortalized in the novel *Grapes of Wrath*, a masterpiece by Nobel prize-winning author John Steinbeck, and it was etched into the minds of Americans by one of the most famous photographs ever taken, *Migrant Mother* by Dorothea Lange.[3] These portrayals of migrant farm workers moving from Dust Bowl Oklahoma to California in the Great Depression vividly illustrate internal migration in response to agricultural labor-market disequilibria as well as the peculiarity of farm labor demand. Dust Bowl migrants discovered that commercial farms hired seasonal crews, not year-round "Jeffersonian" farm hands.

The influx of domestic "Dustbowl workers" into the farm labor market is illustrated in Fig. 6.3. The share of immigrant workers decreases in response to this large, short-lived migration of American farm workers. Prior to the Dustbowl, most farm workers were immigrants (A), but during the Dustbowl thousands of domestic workers crowded immigrants out of the farm labor market (B). Even though the workforce was mostly American, wages remained low because the domestic workers were willing to work at much lower wages than they previously would have. Consequently, the domestic farm labor supply during the Dustbowl became more elastic, as illustrated by a flatter labor supply curve, LS_{dom}' in panel (B) of Fig. 6.3.

For the first time, Americans observed thousands of other White Americans and their families, not so unlike themselves, living in abject poverty, working long days, and earning meager wages to bring food from the fields to the market where the rest of America would consume it. Some of the conditions endured by Dustbowl migrants were captured and documented by academics and journalists as well as authors and photographers like Steinbeck and Lange.

---

2. *A number of striking short videos on the Dustbowl are available; for example, see the History website,* http://www.history.com/topics/dust-bowl/videos/black-blizzard.

3. See US Library of Congress, Prints and Photographs Reading Room; https://www.loc.gov/rr/print/list/128_migm.html (Accessed 22 April 2016).

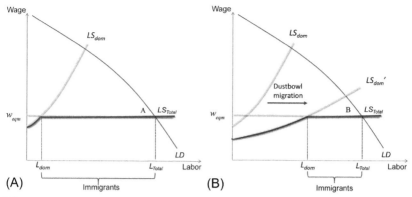

FIG. 6.3   The impact of the Dustbowl migration on the Western US farm labor market. (A) The farm labor market before the Dustbowl. (B) The farm labor market after the Dustbowl migration. The dustbowl resulted in a large rise in domestic farmworkers and a decline in immigrant workers, though wages remained low.

Impoverished farmworkers suddenly were on Americans' radar screen. How could a system of large-scale agriculture be allowed to keep a migrant workforce—of Americans, "like you and me"—in poverty, employ them seasonally at low wages, and have no regard for their living conditions? The federal government stepped in, providing support to 77,118 "transients" in California in 1935 (McWilliams, 1939). UC Berkeley agricultural economist Paul Taylor's research provided key evidence to the 1939 La Follette Committee hearings in the US Senate on violations of civil liberties of farm workers. Carey McWilliams, a lawyer and writer, documented the history and discrimination associated with large-scale farming in the United States and specifically California in his bestselling book *Factories in the Field* (1939).

These images of American farmworkers eventually caught the attention of federal policy makers. Dorothea Lange's photographs motivated the US government to rush aid to migrant camps to prevent families from starving. Upton Sinclair, author of *The Jungle* (1906), ran for governor of California in 1934, staging his platform around a program to "End Poverty in California" and turn large farms into cooperatives. Although Sinclair did not win the election, his ideals and presence in the public spotlight demonstrated a shift in public awareness and attention to the plight of agricultural workers.

The destitute conditions of agricultural workers from the Dustbowl raised concerns that migrant workers would become targets of Communist organizers (Martin, 2009). Farmers feared that workers would organize and strike. These fears led to numerous raids on Hooverville camps by law enforcement authorities and vigilantes citing the need to stop Communist uprisings (Taylor, 2010). Eventually, Californians began guarding the state border with guns, preventing additional Dustbowl migrants from entering (and, in the process, violating the US Constitution's guarantee of the right of free movement across state lines). Despite California's efforts to control it, the Dustbowl represented a massive migration in US history and a turning point in California farm labor history.

Meanwhile, California agriculture continued to expand, as new opportunities for agricultural diversification appeared. The federally sponsored California Central Valley Project, devised in 1933, channeled snow melt from the northern Sierra Nevada mountains into a complex network of reservoirs and aqueducts. It eventually would become a global model for harnessing water in one place (water-rich northern California) and transporting it to another (Southern California and the Central Valley).[4]

WAR CHANGES EVERYTHING

World War II changed everything, as the war effort called able-bodied men and women across the country to other duties, from fighting in the European and Pacific theaters to manning (and "womaning;" see Rosie the Riveter[5]) factories. The domestic supply of farm labor rotated inward again, as domestic workers began to vanish from the fields. The threat of worker uprisings diminished, and farmers urgently called upon the government to give them access to a new source of foreign workers. Agriculture was of strategic importance to feed America and also its troops. Distracted by war, the American public turned its attention away from the conditions and needs of farmworkers.

THE SECOND BRACERO PROGRAM (1942–64)

Following the Dustbowl migration, Mexican immigrants dominated California's agricultural landscape. In 1942, Presidents Franklyn D. Roosevelt (United States) and Manuel Camacho (Mexico) signed an agreement for a new guest worker program between Mexico and the United States. This was the second Bracero Program. Its goal was to support the war effort by filling alleged labor shortages on US farms.

The Mexican government demanded assurances that this would be better for Mexicans than the original Bracero Program had been. Bracero workers were required to work at least 75% of the days in their contract. Mexico stipulated additional requirements to prevent Bracero workers from accumulating debts because of farmers overcharging for room, board, or transportation to the fields, as many had done during the first Bracero Program. Farmers were to make contracts with the workers, provide housing and transportation from Mexico at no

---

4. *California Department of Water Resources*, "California State Water Project Today," Retrieved 4-22-2016. http://www.water.ca.gov/swp/swptoday.cfm.

5. https://www.history.com/topics/world-war-ii/rosie-the-riveter, Accessed 21 May 2018.

cost to the worker, provide meals or cooking facilities, provide worker's compensation insurance, and pay wages at least as high as those of American workers in comparable jobs. Nevertheless, Mexican farmworker wages remained low.

Many Americans opposed the Bracero Program, convinced that farmers were using it to keep farmworker wages low. American labor unions were strictly opposed the program, claiming that Americans would accept jobs as field workers if decent wages and working conditions were guaranteed (Craig, 1971). Although continuation of the program was debated in Congress, grower interests were well represented, and legislation was passed to support the hiring of Mexican workers, as Braceros or without authorization. Unauthorized workers could be deported, but employers were not prosecuted for hiring illegal immigrants.

The Bracero Program did not end when the war ended in 1945. Between 1942 and 1964 more than 4.5 million Bracero workers came to work in the United States But the program came to an abrupt end in 1964. On September 17, 1963, a bus carrying Bracero workers home after a 10-hour shift of harvesting celery and other vegetables collided with a train in Chualar, in the Salinas Valley of California. The bus was an illegally converted flatbed truck. In all, 32 Braceros were killed and 27 were injured. The incident drew public attention, and policy-makers became concerned about the unsafe conditions allegedly perpetuated by the Bracero Program and the exploitation of Mexican workers. Debates in Congress about whether to continue the Bracero Program were renewed, the central argument for ending the Bracero Program being that immigrants to US farms illegally depressed farmworker wages.

The Bracero program had a far-reaching impact on US agricultural production. California fruit and nut output rose 15% during the 1950s, and vegetable production rose 50%. However, farm worker wages rose hardly enough to keep up with inflation. The average farmworker wage in California rose 41%, from $0.85 in 1950 to $1.20 in 1960, while the average factory worker wage in the US rose 63%, from $1.60 in 1950 to $2.60 in 1960 (Martin, 2003). The US Consumer Price Index, a key measure of inflation, rose 23% during this period. This means that, in real (inflation-adjusted) terms, farmworker wages increased at a rate of only 1.4% per year, compared with 2.8% for factory workers.[6]

There was no readily available alternative to Mexican workers in 1964. What did this imply for US growers? Fig. 6.4 illustrates the expected impact of a sudden disappearance of Bracero workers. Braceros had provided an elastic supply of labor at wage w (point A in the figure). However, if the immigrant labor supply

---

6. To calculate average annual wage increases (ω), we used the formula $\omega = (Ln(w_{1960}) - Ln(w_{1950}))/10$, which derives from the exponential growth function $w_{1960} = w_{1950}e^{\omega t}$ evaluated at $t = 10$ years. Adjusted for 23% inflation from 1950 to 1960, in 1950 dollars the farm wage was $0.85 in 1950 and $0.97 (= $1.20/1.23) in 1960, and the nonfarm wage was $1.60 in 1950 and $2.11 in 1960.

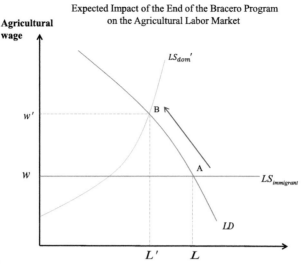

FIG. 6.4 Potential consequences of the end of the Bracero Program. The end of the Bracero Program was expected to shift the farm labor supply inward from the elastic immigrant farm labor supply to the more inelastic domestic supply, raising wages and decreasing the number of farm workers employed. This is shown by a movement from point A to B in the figure.

is removed from the figure, the total labor supply is only the supply of domestic workers to agricultural work, represented by curve LS_{dom}'. A small share of the US workforce worked in agriculture in the 1960s, and the domestic farm labor supply was inelastic compared with the farm labor supply from Mexico. Without an immigrant workforce, the equilibrium agricultural wage in Fig. 6.4 rises to w', and the number of workers employed in agriculture decreases from L to L' (point B). Consequently, domestic agricultural production must decrease as well.

ILLEGAL IMMIGRATION AFTER THE BRACERO PROGRAM

Even as the Bracero Program came to a close, the labor force from rural Mexico was expanding. An unprecedented number of children were growing up in rural Mexico, thanks to one of the highest fertility rates in the world (above seven children per woman in 1960, higher than India's).[7] Rural Mexicans had little access to education, and many could not imagine life without farm work. But they had one thing that no generation before them had ever had: easy access to farm jobs in the United States, jobs that paid low wages by US standards but

---

7. World Bank Indicators. Total Fertility Rate is the estimated average number of children that a woman would have in her lifetime, based on present-day age-specific birth rates and assuming no mortality during the childbearing years.

more than 10 times what an able-bodied person could hope to make doing farm work in a Mexican village (if he could find work in the village).

Villages throughout Central Mexico had been the recruitment center for the Bracero Program. Village children grew up hearing stories of how their fathers, grandfathers, uncles, and neighbors had worked the fields around Imperial, Salinas, Sacramento, even up into Oregon and Washington, following the crops. An unprecedented number had relatives still living and working in the fields of California, Arizona, or Texas; or even farther afield, in Florida, where they picked oranges and tomatoes, or Virginia, where they harvested tobacco, or Pennsylvania, in mushroom greenhouses.

Before cell phones hit the scene, US migrants were only a call away on phones found in village stores all over rural Mexico. They were conduits of news from fields in the United States to villages in Mexico: where the best jobs were, how much they paid, and how to find them. Family migrants became de facto labor recruiters. In December, at the fiesta of the Virgin of Guadalupe, they came home with the vestments of success: nice clothes, gifts for the family, maybe even a car. For a village teenager, they were someone to look up to. For families throughout Central Mexico, they were a reminder that sending their older children north could mean putting better food on the table, a new roof on the house, pouring a cement floor, cash to send a younger sibling to school or to buy a new plow, TV set, or support in old age.

For hundreds of thousands of Mexicans, having a relative or friend in the United States made the difference between migrating and staying home. Without a close contact in the US, migration is expensive and risky. Migrants have to leave the village with enough money to cross the border and survive until they find work. They run the risk of losing everything if the US border patrol catches them crossing the border without papers and sends them back to Mexico. To increase their chances of success, they might pay a human trafficker, or *coyote*, to guide them across the border, but coyotes charge steep rates. Migrants often have to borrow money to cover the cost of the northward trek, perhaps from fellow villagers, but there is no guarantee they will be able to repay their debts. Even with a coyote, the illegal border crossing is never certain. An unscrupulous coyote might steal the migrants' money and abandon them before they cross the border, or worse, migrants could be abandoned, lost, or even wind up dead in the Arizona desert.

With a contact in the United States, the risks, fears, and anxieties of migrating greatly diminish. Rather than raising enough cash to keep the migrant until he or she can find a job, the migrant with contacts may need only sufficient funds to get to the border. Relatives or friends in the United States put the prospective migrants in touch with a trusted coyote, and they pay the coyote once the new migrants arrive safely at their destination in the United States. It is like a loan and an insurance policy bundled into one package. The friends in the United States set the new migrants up with a job, and the new migrants do not have to pay back their friends until after the wages start coming in. If they manage their money carefully, they might even be able to start sending cash (remittances) to their families in the village shortly after they arrive.

Social scientists call these family contacts "family migration networks." They have invested a lot of research effort into documenting how networks drive new migration, making migration a self-perpetuating process: more migration means more networks which, in turn, mean more migration in the future. Sociologists call this *cumulative causation* (Massey et al., 1993).

In a mover-stayer model (Chapters 3 and 4), from the would-be migrant's or his family's point of view, illegal migration is optimal if the benefits associated with migration exceed the costs, including the opportunity cost of migrating legally or not migrating at all and working in Mexico. During the Bracero Program, a system was in place for people to migrate legally to work in the United States. It was not free—some Bracero recruiters made a small fortune selling permits to migrants. As more and more Braceros migrated to the United States, a Bracero migrant network emerged. Braceros were a legal anchor in the United States; they could help other family members migrate illegally, bypassing the Bracero system altogether. Some Braceros did not bother getting permits for their next trip. As networks lowered migration costs, illegal migration expanded alongside Bracero migration during the Bracero period. Even more Mexicans came to work on US farms without authorization than with Bracero visas (Taylor, 2010). Fig. 6.5 shows that, as the number of Bracero workers increased, border apprehensions of unauthorized immigrants surged.[8]

After the Bracero program ended in 1964, many Braceros continued to work in US fields as illegal immigrants. Meanwhile, new unauthorized immigrants came into the United States from Mexico to do farm work. Many aging Braceros became labor recruiters. They understood how migration networks functioned, and they used these networks to recruit workers and supply them to farms. It was easy for recruiters to tap into networks and find workers willing to migrate from poor villages in Mexico to do farm work in the United States. As migrant networks expanded and migration increased, recruiters let networks do much of the work for them. They became a first employer of many migrants after they arrived in the United States.

The US Immigration and Naturalization Service (INS) had the unenviable task of patrolling the 1951-mile Mexico-US border. However, a few INS agents were no match for the number of Mexicans attempting to cross northward, lured by a seemingly insatiable demand for low-skilled workers in agriculture and other sectors of the US economy. One study found that more than two thirds of individuals succeeded in crossing the border illegally on their first try (Massey and Singer, 1995).

---

8. These numbers are not comparable in a strict sense, because the Bracero curve shows numbers of workers, while apprehensions are events—the same worker may be apprehended trying to enter the US more than once. Researchers often use border apprehensions as a proxy or indicator of illegal immigration.

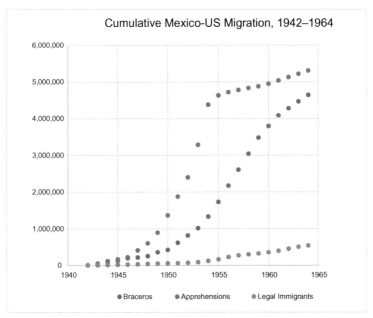

FIG. 6.5   As the number of Bracero workers increased, so did apprehensions of unauthorized immigrants, reflecting a surge in illegal immigration. *(Data from Congressional Research Service, Temporary Worker Programs: Background and Issues, Washington, DC, pp. 36–37, 1980.)*

By the 1980s, more than half of the California farm workforce was unauthorized.[9] By not providing a way for rural Mexicans to work on US farms legally, US immigration policy guaranteed that the US farm workforce would become a largely illegal one.

Agriculture was not alone when it came to relying on unauthorized immigrants. A number of other sectors of the US economy became dependent on illegal immigrants to do low-skilled work, from janitorial services to gardening, construction, and child care. What made agriculture different from those other sectors is that farmers were relatively well organized, admitted to relying on a largely illegal immigrant workforce, and managed to exert influence on lawmakers when a backlash against illegal immigration swept through the country in the 1980s.

Having an economy in which entire sectors depended on unauthorized workers was not popular among many American voters and politicians. In

9. Not surprisingly, reliable data on the number of farmworkers who are unauthorized are not available. Our estimate is based on how many California farmworkers applied for legal status under the Special Agricultural Worker (SAW) program after 1987. That number easily exceeded 50% of the top-end estimate of the number of California farmworkers in that year.

the early 1980s, immigration reforms in the US Congress threatened to end the continued employment of unauthorized workers on US farms.

THE 1986 IMMIGRATION REFORM AND CONTROL ACT

In the 1980s, there was a chorus for immigration reform in the United States. Even the major farmworker union, the United Farmworkers of America (UFW), joined in, arguing that illegal immigrants were obstacles to their organizing efforts. In 1981, the legislative director of the UFW urged Congress to impose sanctions on employers who hired illegal immigrants.[10] The UFW-organized farm workers were mostly foreign born, but high turnover fed by new immigration stifled organizing efforts.

The Immigration Reform and Control Act (IRCA) of 1986 proposed a compromise. It had two major components:

(1) Employer sanctions, or fines and, after repeated offenses, possible imprisonment of employers who knowingly hired illegal immigrants. Previously, it had not been illegal to hire an unauthorized immigrant, even knowingly. This provision was strongly supported by conservatives as well as by the UFW. The goal of this provision was to discourage unauthorized entry and employment in the United States.

(2) Legalization of unauthorized aliens who had resided continuously and long enough to develope an "equity stake" in the United States. This provision, strongly supported by liberal groups and immigrant advocates, recognized that a large number of people had been living in the United States, working, and contributing to the economy and society long enough that they should be given the chance to become legal residents and, eventually, citizens.

Western crop farmers opposed this compromise. They argued that it did not acknowledge their unique dependence on unauthorized workers, and they would not be able to obtain workers in the flexible manner necessary for their "perishable agriculture" via the existing guest worker program, known as the H-2 program.

Western growers demonstrated in 1984–85 that they had the political clout to block immigration reforms that did not deal with their concerns. A compromise was reached in the summer of 1986. Under the so-called Schumer compromise, named for then Representative Charles Schumer (D-NY), unauthorized farm workers would be legalized under the Special Agricultural Worker (SAW) program, giving agriculture a legal work force. If these legalized workers left farm work quickly and farm labor shortages developed, additional workers could be admitted via the revised guest worker (H-2A) program or via the Replenishment Agricultural Worker (RAW) program.

---

10. Testimony by Stephanie Bower before the US Senate Subcommittee on Immigration and Refugee Policy, "The Knowing Employment of Illegal Immigrants," Serial J-97-61, September 39, 1981, p. 78. (Cited in Martin, 2009).

There was a fundamental difference between the H-2A and RAW guest worker programs. Farmers had to attempt to recruit workers under US Department of Labor (DOL) supervision to obtain H-2A workers, but they were rewarded with workers tied to their farms by contracts. Under the RAW program, by contrast, a certain number of foreign workers would be admitted to "float" from farm to farm, so that farmers would not have to engage in DOL-supervised recruitment or provide housing to RAW workers. RAW workers could earn an immigrant status by continuing to do farm work.

There were no farm labor shortages in the early 1990s, and the RAW program was allowed to expire without ever being used. Instead, continued unauthorized entries and employment led to surpluses of workers and falling wages. The average hourly earnings of farm workers fell relative to average manufacturing wages in the early 1990s, despite a recession and escalating health care costs that held nonfarm wages in check (Fig. 6.6). This prompted the US Commission on Agricultural Workers (CAW) to conclude that there was "a general oversupply of farm labor nationwide" and, "with fraudulent documents easily available," employer sanctions were not deterring the entry or employment of unauthorized workers (CAW, 1993, pp. xix–xx).

The SAW guidelines were written so that, once a worker claimed to have done the requisite number of days of farm work, the burden shifted to the US government to disprove the claim. In most cases, a simple affidavit from a farm employer stating that an individual had worked 90 days or more in perishable crops between May 1984 and May 1985 was sufficient to become legalized under the SAW program. The burden of proof was on the government to

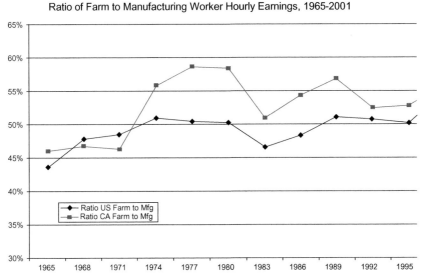

FIG. 6.6 Ratio of farm to manufacturing hourly earnings, 1965–95.

show that the worker had not actually performed the requisite days of farm work. Because of this, the SAW program became riddled with fraud and an easy back-door to legalization.

Some government interviewers conscientiously tried to make sure that a SAW applicant's claim was legitimate. For example:

> There was a conscientious first-line interviewer who knew something about the rural life and who faced many SAW applicants who did not. She developed a loose-leaf notebook with no text. It consisted of pictures of, and dried leaves from, the kinds of crops that were grown in her area. If someone said that they had picked strawberries, she asked them to show her the strawberry plants in the book; if they could not properly identify them she recommended a denial.[11]

There were some cases in which the government succeeded in proving that a worker's SAW application was fraudulent:

> There were countless anecdotes of fur-coat wearing Europeans seeking SAW status in Manhattan, applicants who contended that the cotton they harvested was purple, or that cherries were dug out of the ground, or that one used a ladder to pick strawberries. Often the temporary INS staffers handling the SAW applications were as clueless about agriculture as some of the applicants."[12]

Martin, Taylor, and Hardiman's (1988) analysis of California Employment Development Department data found that the 1.2 million approved SAW workers substantially exceeded the number of individuals who actually worked the requisite 90 days, suggesting widespread fraud.

Newly arrived unauthorized foreigners soon became a significant majority of the hired farm work force, and increasingly they were brought to farms by farm labor contractors (FLCs) willing to be "risk absorbers" in the event of immigration law enforcement. One study found that California farmers' use of FLCs increased sharply in the wake of IRCA (Taylor and Thilmany, 1993). In general, the FLC, not the farmer who contracts him, is the employer of record under US immigration law. It is exceedingly difficult to prove that a farmer knowingly hired illegal immigrants through a FLC. It is also difficult to demonstrate that a farmer had knowledge of FLC violations of labor laws. Thus, FLCs provide farmers not only with short-term labor, but also with a buffer from immigration and labor laws. Evading these laws is a comparative advantage of some FLCs. FLCs also can help workers skirt immigration laws, for example, by providing them with fictitious documents and rides to the fields, and their mobility can make them difficult to regulate.

---

11. David North, "A Bailout for Illegal Immigrants? Lessons from the Implementation of the 1986 IRCA Amnesty," Washington, DC: Center for Immigration Studies; http://cis.org/irca-amnesty.
12. David North, "Lessons Learned From the Legalization Programs of The 1980s," Immigration Daily; http://www.ilw.com/articles/2005,0302-north.shtm.

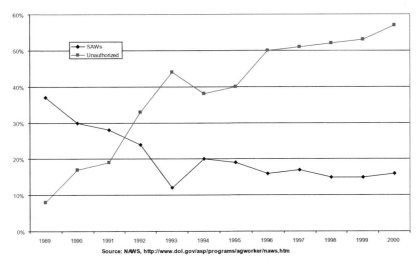

FIG. 6.7 SAWs and unauthorized crop workers: 1989–2000.

Many legalized SAWs moved to cities to find jobs in services, construction, and manufacturing. By 1993, the SAW share of the US crop work force had fallen by two-thirds and the unauthorized share had quadrupled to nearly 60% (U.S. Dept. Labor, 2005; Fig. 6.7). A *Wall Street Journal* editorial noted that migrants, "often illegal," had created an "efficient... informal guest worker program." Farm employers worried about losing crops if there were effective immigration enforcement. One analyst wrote: "We have an agricultural industry in this country that is totally dependent ... your investment, your livelihood, my food supply, our national food security are totally dependent on illegal workers."[13]

The National Agricultural Worker Survey (NAWS) of 2001–02 found that 16% of the US hired farm workforce were newcomers, and 99% of newcomers doing farm work in the United States were unauthorized. (Newcomers are defined as people who have been in the United States 1 year or less.) At this recharge rate, it does not take very long for the farm workforce to become overwhelmingly unauthorized again. In 2005–06, less than two decades after the SAW legalization, an estimated 73% of California's farm work force was unauthorized.[14]

UNINTENDED CONSEQUENCES: THE IMPACTS OF IRCA ON US AGRICULTURE

IRCA had three major impacts on the farm labor market.

---

13. Jim Holt, expert and consultant on the H-2A program, cited in AgAlert, August 8, 2007; http://www.agalert.com/story/?id=872.
14. Estimated from NAWS Public Use Data (http://www.doleta.gov/agworker/naws.cfm).

First, it created a new breed of documented illegal immigrant workers. Under IRCA's rules, employers are required to fill out a form, called the "I-9" form, certifying that the workers they hire possess the documents necessary to "prove" their legal status. For this, various documents or combinations of documents were deemed acceptable under the law. For example, a Social Security Card and driver's license would be sufficient to comply with the I-9 requirements. Both became available for purchase at flea markets and other locations around the country. Employers were not required to verify the authenticity of the documents their workers showed them. This was a major loophole in the law.

Second, IRCA created incentives for farmers to avoid hiring workers directly, and instead strike contracts with FLCs. If a farmer contracted with a FLC to harvest a crop, the FLC generally became the employer of record under IRCA, responsible for complying with I-9 requirements. FLC operations are notoriously difficult to monitor when it comes to immigration and labor laws. With their keen knowledge of migration networks, FLCs had a comparative advantage in quickly assembling harvest crews of largely illegal immigrant workers, taking them out to farms, then disbanding their crews once they were no longer needed. In theory, farmers could be fined for knowingly hiring unauthorized workers through a FLC, but in practice it was virtually impossible to prove that farmers knew the FLC's workers were illegal. By using FLCs, farmers could effectively buffer themselves from IRCA's employer sanctions.

Third, paradoxically, IRCA increased the supply of agricultural workers. The SAW legalization program was not just successful at legalizing the farm workforce; it was *too* successful. The SAW legalization program conveyed the message to potential migrants that it was easy to obtain legal status in the United States if one stayed there long enough as an illegal immigrant.

Although IRCA had many unintended consequences, it was not effective at reducing employment of illegal workers. Few fines were levied against farmers for knowingly hiring illegal immigrants. This was partly because the INS dedicated few human resources to enforce immigration laws inside US borders and at the workplace. It was also because of how difficult it was for the government to prove that a farmer *knowingly* hired illegal immigrants. If the farmer could produce completed I-9 forms for her workers, the burden was on the government to prove that the documents workers provided were bogus *and* that the farmer was aware of it. As farmers shifted to using FLCs instead of hiring workers directly, it became even harder for the government to enforce IRCA's employment sanctions.

The SAW program effectively legalized the farm workforce, but IRCA did not slow the flow of new unauthorized workers into the United States, for two reasons: First, there was abundant supply of potential farmworkers in rural Mexico, fed by high birthrates, low education, low wages, and high unemployment in rural Mexico; and second, migration networks connected potential migrants in rural Mexico with farm jobs in the United States.

There might be a third reason why IRCA did not slow unauthorized immigration. The SAW program conveyed to rural Mexicans that they might eventually be able to qualify for legalization under a future SAW program if they migrated illegally to the United States soon. Not only that, once immigrants become legalized, family reunification provisions in US immigration law create an "immigration multiplier." One study found that that each additional legal immigrant admitted to the United States from Latin America results in a 3.25-person increase in immigration. This includes the legalized immigrant, plus an additional 2.25 via family reunification.[15] As a result of the SAW program, more rural Mexicans had *legal* family contacts in the United States than ever before.

IMPORTING LEGAL FARM WORKERS (AT A COST): THE H-2A PROGRAM

There is one way to import workers legally to work on US farms. Under the H-2A program, farm employers can temporarily bring workers into the country for temporary or seasonal work lasting 10 months or less. H2 temporary guest worker visas have existed since 1953. In 1986, IRCA subdivided this category into two: the H-2A sub-category of temporary agricultural workers, which was not limited or "capped," and the H-2B category of temporary workers in other sectors, which was capped. By not capping the number of H-2A workers, IRCA recognized the need to counter a decline in illegal immigration that was expected to happen once the US government increased border security and immigration enforcement.

H-2A workers are "temporary nonimmigrant" workers, permitted inside the United States for only a limited period of time to be employed exclusively by the employer who requests their visa. Once their work season is finished, they are required to return home. In order to obtain H-2A workers, farm employers have to petition the US DOL and show that:

(a) There are not enough US workers "able, willing, qualified, and available to perform work at the place and time needed," and

(b) The wages and working conditions of workers in the United States will not be "adversely affected" by the importation of guest workers.

US Citizenship and Immigration Services have to approve the employer's request for H-2 workers, and then the workers have to apply for their H-2 visa at a US embassy or consulate abroad. H-2 workers have to be provided with free housing. Workers who complete half the season must be reimbursed for transportation costs to the place of employment, and those who complete the full

---

15. See Yu (2005). An additional 1.78 people are added to the US population from these immigrants giving birth to second-generation immigrants. Because they are born in the United States, these children of immigrants are US citizens.

season also must be reimbursed for transportation costs of returning home. Other rules and protections apply. These requirements make using H-2A workers more expensive than hiring local workers.[16]

Farmers in some regions, particularly tobacco growers in the southeastern states, turned to the H-2A program to obtain legal foreign workers. Growers in the West, however, argued that the H-2A's requirements were too cumbersome and out of step with the needs of FVH production, where labor demands are variable and timing is critical. In perishable crops, workers have to be available at the right place and time.

Not all analysts agreed. Jim Holt, a farm labor expert who assists farmers with obtaining H-2A workers, claims: "There is folklore that somehow or another the H-2A program is not available in California or does not work in California. That is simply not the case. H-2A has always been an option here in California, but not so much an attractive option."

Regardless of whether an H-2A program could work for FVH farms, in an environment with abundant unauthorized labor and easy access to this labor through contractors, there was little incentive for most farmers to incur the time and money costs of recruiting H-2A workers. In 2006, there were only 59,112 H-2A workers in the United States, out of a total hired farm workforce of more than 2.4 million. California had 2292 H-2A workers, out of a total hired farm workforce of about 800,000.[17] In both cases, most of the rest of the hired farm workforce was unauthorized just before IRCA legalized most of the farm workforce, and it became mostly illegal again by the end of the century. This number has risen in recent years, for reasons discussed in Chapters 8 and 9, but it remains small relative to the total hired farm workforce and the immigrant workforce.

EFFORTS AT IMMIGRATION AND FARM LABOR REFORM AFTER IRCA

Recognizing that enforcement could leave them without harvest workers, but still unwilling to undergo DOL-supervised recruitment and satisfy the housing requirements of the H-2A program, western growers proposed variations of the expired RAW program to Congress. There was widespread union, church, and Hispanic advocacy opposition, and in 1995 US President Clinton threatened to veto any such program: "I oppose efforts in this Congress to institute a new guest worker or '*bracero*' program that seeks to bring thousands of foreign

---

16. A helpful summary can be found at the U.S. Department of Labor website: http://www.dol.gov/whd/regs/compliance/whdfs26.htm.

For a critique of the H-2A Program and recent changes, see: https://www.farmworkerjustice.org/content/h-2a-guestworker-program.

17. Estimates by Philip L. Martin, reflecting that an average of two workers fills every full-time equivalent farm job. See "COA: Jobs to Workers (2006).

workers into the United States to provide temporary farm labor" (White House Press Release, June 23, 1995).

Farmers did not give up on their quest for an alternative guest worker program. Following the US and Mexican presidential elections of 2000, farm employer and worker advocates agreed on a compromise that introduced a new concept: earned legalization. Earned legalization would offer temporary resident status (TRS) to unauthorized workers who had done at least 100 days of farm work during the previous year and allow them to earn an immigrant visa if they did at least 360 more days of farm work in the next 6 years. Spouses and minor children of TRS workers would not be deportable, but they would not be allowed to work in the United States. They could receive permanent immigrant status once the farm worker received an immigrant visa, regardless of queues and waiting lists. Earned legalization satisfied employers, who received assurance that newly legalized farm workers would not immediately leave for non-farm jobs. It also satisfied worker advocates, who wanted legal immigrant status for unauthorized farm workers.

During the spring and summer of 2001, there were binational Mexico-US meetings on migration, which had become Mexico's top foreign policy priority under President Vicente Fox. A variety of proposals were introduced in the US Congress to legalize foreign-born workers. The debate centered largely on whether currently unauthorized workers in the United States should be granted guest-worker status, immigrant status, or a TRS that would enable them to eventually "earn" legal immigrant status. The September 11, 2001 terrorist attacks in the United States stopped legislative momentum for these proposals, turning the US government's attention to border security, instead.

Earned legalization was one of two central provisions of the Agricultural Job Opportunities, Benefits, and Security Act, or AgJOBS. However, many Congressional Republicans opposed it because it "rewarded lawbreakers" with legal status.

AgJOBS also included an effort to make the H-2A program more "employer friendly." Instead of having to demonstrate that there was a shortage of US workers and that H-2A workers would not adversely affect US workers, farmers could simply "attest" to their need for foreign workers. Farmers seeking H-2A workers would have to file job offers at local Employment Service offices at least 28 days before the workers were needed. Employers would also have to advertise the jobs in local media at least 14 days before the need date. If local workers did not appear, the US DOL would be required to authorize employers to employ H-2A workers.

Employers would be allowed to provide H-2A workers with either housing or "a monetary housing allowance," provided the state's governor certified that there was sufficient housing for workers to find their own. The housing allowance would be the "statewide average fair market rental for existing housing for metropolitan or nonmetropolitan counties," assuming two persons per bedroom in a two-bedroom unit.

Employers of H-2A workers would have to reimburse inbound and return transportation costs for satisfactory workers and guarantee work for at least three quarters of the period of employment that they specified. In a bow to worker advocates, H-2A workers would be able to sue in federal rather than state courts to enforce their contracts. Housing and other H-2A provisions could be superseded by a collective bargaining agreement.

Farmers would have to pay H-2A workers the higher of the federal or state minimum wage, the prevailing wage in the occupation and area of intended employment, or the adverse effect wage rate (AEWR). US citizen, legal immigrant, and unauthorized workers could be paid the federal or state minimum wage unless they worked alongside H-2A workers. To win grower support for AgJOBS, the lower 2002 AEWRs would apply until 2006. The 2002 hourly AEWR was $8.02 in California, $7.69 in Florida, $7.53 in North Carolina, $7.28 in Texas, and $8.60 in Washington.

These attempts at reform illustrate how complicated it is to solve the farm labor problem via government policies. After 2000, AgJOBS was the most serious US congressional effort to solve agriculture's reliance on unauthorized immigrants while ensuring farmers' access to foreign workers. One can imagine the extensive negotiations and effort that went into crafting all of its provisions, which sought to satisfy both farmers and farm worker advocates. The AgJOBS bill was introduced in Congress in 2007 and again in 2009 but died for lack of support in both cases, testimony to the controversial and politically charged nature of immigration reform in the United States.

GUEST WORKER PROGRAMS IN OTHER COUNTRIES

Farm labor immigration policies range from tight government regulation of foreign recruitment and labor management, to not-so-benign neglect of unauthorized immigration, to regimes of limited enforcement punctuated by an occasional amnesty program.

New Zealand's Recognized Seasonal Employers (RSEs) illustrates the extreme of government-regulated recruitment of foreign workers. Employers must try to recruit local workers before obtaining permission to import guest workers from the Pacific Islands, and once they receive approval, they are required to abide by strict wage, travel reimbursement, housing, and other regulations. The system is not without controversy. The Chair of the organization representing New Zealand's pip fruit (apple and pear) industry warned in 2009 that "the scheme will not meet next season's labor requirements, and there need to be alternatives on offer to help with the transition...growers will rely on illegal labor if that's not given."[18] Nevertheless, economists Gibson and McKenzie (2014) hold this program up as a model to other guest worker programs. RSE was designed with the intention of promoting economic development in the

---

18. *Daily New Zealand News*, Tuesday, July 31, "Illegal workers tipped in orchards–Pipfruit NZ.

Pacific Islands, and in fact, Gibson and McKenzie find that households in Tonga and Vanuatu that participated in the RSE program earned 35% more income on average over a 2-year period. The RSE influenced the design of Australia's Pacific Seasonal Worker's Pilot (PSWP) program.

Canada's Seasonal Agricultural Worker Program (SWAP) requires that participating farmers offer temporary foreign workers a minimum of 240 hours of work within a period of 6 weeks or less for a maximum duration of 8 months between January 1 and December 15. Most SWAP workers are from Mexico, and there appears to be an increasing number of unauthorized Mexican immigrants working on Canadian farms.

Greek agriculture relies significantly on Albanian workers, and its policies illustrate a more adaptive and laissez-faire approach.[19] Many Albanians entered as illegal migrants following the collapse of Albania's communist government in 1991 and a financial crisis provoked by pyramid schemes in the banking sector in 1996. The Albanian exodus strained relations between the two countries, which were exacerbated by a lack of concrete immigration policies in either country. Two regularization schemes, the first in 1997 and the second in 2001, awarded legal residence and work permits to illegal immigrants who had been in Greece for over a year and could provide proof of employment, not unlike the 1986 US amnesty program.

Spain exemplifies an accommodative approach to agricultural labor migration. Farmers seeking guest workers make their requests to provincial labor authorities in the fall, who send them to the *Dirección General de Inmigración* (DGI) in Madrid (Rural Migration News, 2008; Migration News, 2009). The DGI announces the *contingente* or quota by December for the following year. Agricultural guest workers receive seasonal permits for 3–9 months that must be renewed outside Spain; there is no cap on the number available. There are relatively few illegal workers, in part because employer organizations are able to obtain sufficient guest worker permits and because sanctions range from €6,000 to €60,000 for each illegal hire.

European integration and the breakup of the former Soviet Union opened up new sources of farm labor for Western Europe. Under the Schengen accord, members of European Union (EU) countries may work freely in any other EU country. As an EU member, the United Kingdom (UK) permits workers from the A8 Countries of Eastern Europe to travel freely to the United Kingdom and work after registering with the Workers Registration Scheme.[20] Farm labor demand during the peak season in Britain is estimated at about 4.5 times the demand during the low season (Migration Advisory Committee, 2017). Few of these jobs are filled by domestic workers. After the United Kingdom's exit from the EU (Brexit), the UK faces the prospect of severe farm labor shortages.

---

19. See *Central Europe Review* (1999).
20. The A8 countries include the Czech Republic, Estonia, Hungary, Latvia, Lithuania, Poland, Slovakia, and Slovenia (Rural Migration News, 2009).

Andrew McCornick, the National Farmers Union (NFU) Scotland President, estimated a seasonal farm labor shortage from the EU of between 10% and 20% in the year following the passage of Brexit. One farmer who produceed potatoes, carrots, parsnips, broccoli, and cauliflower estimated that he left crops valued from 30,000 to 50,000 British pounds unharvested due to labor shortages in 2017 after a fall in the value of the British pound (Agerholm, 2018). Consequent to farm labor shortages throughout Britain, the NFU began advocating for the return of the Seasonal Agricultural Workers' Scheme (SAWS), which permitted agricultural guest workers to come to the United Kingdom seasonally prior to 2013.

As commercial agriculture expands, a widening diversity of countries grapple with the farm labor problem. The importation of agricultural workers is increasingly commonplace, particularly when countries with different levels of economic development share borders.

In September 2009, a recruitment center was opened at the Beitbridge crossing at the Limpopo River to enable South African farmers to recruit Zimbabwean workers (Rural Migration News, 2004; Rural Migration News, 2009). Israel in 2007 increased the number of foreign farm workers to 29,000 despite protests from Ma'an, the Workers Advice Center, which claimed foreign workers would take jobs away from Arab agricultural laborers (Rural Migration News, 2007).

Haitians constitute over 90% of the seasonal sugar work force and two-thirds of coffee workers in the Dominican Republic (Martin et al., 2002). At least 32% of total agricultural wages in the country are paid to Haitians (Filipski and Taylor, 2009).

Costa Rica has a long history of importing farmworkers from its poorer Central American neighbor, Nicaragua. There were an estimated 135,579 non-naturalized Nicaraguans in Costa Rica in 2000, up from 78,487 in 1998. Recent decades have witnessed frequent legal reforms including temporary immigration measures and legalization of some Nicaraguan immigrants. Caitlin Fouratt, an anthropologist, writes that frequent changes in immigration policy and a high degree of bureaucracy "create a sense of Costa Rican immigration law as temporary. The ongoing temporary character of law and the forms of immigrant illegality it generates create uncertainty about the boundaries between legality and illegality among migrants in Costa Rica" (Fouratt, 2016). One in four Nicaraguan immigrant workers in Costa Rica works in agriculture (International Organization for Migration, 2001). Sri Lanka is a major exporter of domestic helpers and other workers; however, it imports Indian workers to harvest tea.

Mexico is unique in being a major country of farm labor emigration, internal migration, and immigration. It is by far the world's largest nation of farm labor emigration, supplying most of the US hired farm work force. Significant internal migration supplies labor to agribusinesses in Mexico, largely producers for export markets in the northwestern states of Sinaloa, Sonora, Baja California and Baja California Sur. The migrant workers are mostly from Mexico's poorer southern states of Oaxaca and Guerrero.

Mexico also imports agricultural workers, principally to fill low-wage jobs in the South. Under a special federal program, Mexico's southernmost state of Chiapas, which is poor, draws temporary agricultural migrant workers from Guatemala, which is poorer.[21] Agricultural wages drop sharply as one moves southward along migratory chains in North America, from an average of US$10.56 an hour in 2016 for field workers in California to US$1.25 in Mexico's northwest region. In Guatemala, the official minimum agricultural wage was US$1.27 in 2015, but by one estimate 90% of Guatemalan farm workers get paid less than the minimum wage. Few agricultural workers have anything resembling year-round employment.[22]

In recent decades, a significant share of foreigners entering Mexico to work in agriculture did so without the formal authorization from the Secretariat of the Interior. In October 1997, as a result of the 5th Mexico-Guatemala Binational Meeting on Migratory Issues, the Mexican government agreed to establish procedures for the documentation of Guatemalan nationals temporarily residing in Mexico to allow them to perform agricultural labor in the southeastern state of Chiapas. The fact that many Guatemalans enter Mexico to do farm work in Chiapas tells a lot about working conditions in Guatemalan agriculture, because working conditions in Chiapas are poor. According to the UC Davis—Colegio de Mexico 2008 Mexico National Rural Household Survey, the average agricultural worker in Chiapas earned $0.98 per hour.[23]

IMMIGRATION AND THE DOMESTIC FARM LABOR SUPPLY: CHICKEN OR EGG?

Do guest workers and unauthorized immigrants drive domestic workers out of the fields? Or do they take farm jobs that domestic workers shun? There is no way to replay the past and see what would have happened without foreign agricultural workers in today's relatively high-income countries. However, there is

---

21. Instituto Nacional de Migración Circular No. CRE – 247-97, "para trabajar temporalmente en las fincas cañeras, ganaderas y plataneras del Estado de Chiapas" (http://www.gobernacion.gob.mx/archnov/MANUALm.pdf). These workers are allowed multiple entries and exits into Mexico, but their movement is limited to within the state of Chiapas. See also Protection of Migrant Agricultural Workers in Canada, Mexico and the United States *Commission for Labor Cooperation,* Secretariat of the Commission for Labor Cooperation (International organization created under the North American Agreement for Labor Cooperation, http://www.naalc.org/english/pdf/study4.pdf).

22. Data on California farm worker wages are from USDA, National Agricultural Statistics Service (NASS), 2014. Mexican wages were compiled from the 2008 Mexico National Rural Household Survey (ENHRUM) and Richard Marosi (2015); http://www.latimes.com/local/california/la-me-baja-farmworkers-20150326-story.html. The Guatemalan minimum agricultural wage is from wageindicator; http://www.wageindicator.org/main/salary/minimum-wage/guatemala.

23. You can learn more about the Mexico National Rural Household Survey at: http://precesam.colmex.mx/ENHRUM_English.html.

strong anecdotal and quasiexperimental evidence that, at least the way farm work looks today, most domestic workers would not have any part of it.

Under the US government's H-2A program, farmers can get permission from the federal government to bring temporary workers into the country, but they must first prove that they have actively recruited US workers and that US workers will not take the jobs they offer. These requirements exist along with the AEWR to minimize any negative labor market impacts. The North Carolina Growers' Association (NCGA) is one of the largest employers of H-2A workers. In 2011, 489,095 North Carolinians were unemployed, only 268 of whom showed interest in doing farm work for the NCGA. Nearly all of them—245 people—were hired, 163 actually showed up for work, but only seven made it through the crop season.

An October 5, 2011 *New York Times* article featured a Colorado corn and onion farmer, John Harold, who usually hired 90 H-2A workers from Mexico each year between May and October (Johnson, 2011). In 2011, with nearly 1 in 11 domestic workers unemployed, he reduced his H-2A workforce by a third, hoping to fill the gap with unemployed domestic workers. His plan backfired. "Six hours was enough, between the 6 a.m. start time and noon lunch break, for the first wave of local workers to quit. Some simply never came back and gave no reason. Twenty-five of them said specifically, according to farm records, that the work was too hard." The article concluded:

> *The H2A program…strongly encourages farmers to hire locally if they can, with a requirement that they advertise in at least three states. That forces participants to take huge risks in guessing where a moving target might land—how many locals, how many foreigners—often with an entire season's revenue at stake. Survival, not civic virtue, drives the equation.*

In 2010, the UFW could not find domestic workers to work in the fields at a time when 14.6 million Americans were unemployed. It launched the "Take Our Jobs" campaign, inviting Americans to take farm workers' jobs.[24] There were few takers. The highlight of the campaign was the appearance of UFW President Arturo Rodriguez on the American television show, The Colbert Report.[25] When asked "How many people of everyone in America have taken you up on this?" Rodriguez responded: "Only three right now are working in the fields." In a later episode, Stephen Colbert attempted to be a farm worker but failed comically.

These examples beg the question of what the farm labor market would look like if there were no immigrants to compete with domestic workers. Do immigrants take jobs that domestic workers would otherwise desire if wages were permitted to rise? It is clear from Fig. 6.1 that immigrants do keep agricultural

---

24. You can learn about this campaign at http://www.ufw.org/toj_play/TOJNEW_12_JAL.html, Accessed 21 May 2018.
25. http://thecolbertreport.cc.com/videos/nlmfgk/arturo-rodriguez, Accessed 21 May 2018.

wages low, but whether domestic workers would be willing to do field labor if wages were permitted to rise depends on how inelastic LS_{dom}' is. You might ask yourself how high the wage would need to be to draw you or your friends into the field to harvest grapes in the 100-degree heat of California's Central Valley.

Germany put this question to the test in a national experiment. It launched a program with the goal of insuring that 10% of all farm workers in Germany were German, subsidizing their farm wages by E13–E20 ($17–$25) tax free. Farmers claim native-born pickers are only half as efficient as immigrant workers from Poland: "Germans lack both practice and motivation," one was quoted as saying. The German government abandoned the program shortly after it was launched, because it was expensive and proved ineffective at recruiting German workers (Tzortzis, 2006).

Some workers and policy makers argue that a large immigrant farm workforce takes jobs from domestic workers by keeping wages unreasonably low. However, there is no clear counterfactual to immigrant farm labor in US history to test this assertion. Jason Resnick, Vice President and General Counsel for Western Growers' Association, explained (Kitroeff, 2017):

> *The one constant is that no matter how much we pay, domestic workers are not applying for these jobs. Raising wages only serves to cannibalize from the existing workforce; it does nothing to add new laborers to the pool.*

IMMIGRATION POLICY AND AGRICULTURE: SOME FINAL THOUGHTS

As this brief history of immigration and farm labor illustrates, policies governing the immigration of farm labor have varied across countries and over time. In the United States, cycles of relatively open immigration policies and low agricultural wages have been followed by rising worker demands, increased fears among domestic workers, discrimination, and policies to ban foreign workers from the fields.

Waves of immigrants not only keep wages low; they also kept land values high:

> *Economists noted that farmers were really using continued immigration and the resulting low wages to protect the value of their land—land 'had been capitalized on the basis of five decades of cheap labor' and high land prices could be 'maintained only with the continued availability of Mexican labor'*

(Fuller, 1940, cited by Martin, 2003, 43).

If farmworker wages had risen as quickly as wages in other industries, the profitability of farming large parcels of land would have decreased along with the willingness to pay for agricultural land—that is, unless a labor substitute were found. Landowners who buy land when farmworker wages are low have incentives to keep wages low. Low farmworker wages maintain high profit margins for landowners while protecting their investment in the land. If new, low-wage immigrants were not available to fill the excess demand for farm labor, farmers would have to respond by finding new sources of cheap immigrant labor, changing their crop mixes, or adopting new technologies to produce food with fewer workers.

In the next few chapters, we shall see that concerted efforts by farmworker advocates and organizers have had some success at changing farm labor laws to improve earnings and working conditions, but progress has been very uneven across countries and US states. As this book goes to press, farmers in the United States and other high-income countries continue to pressure their governments to open the borders to new immigrant farmworkers, while others push to build walls and keep immigrants out. Meanwhile, new technologies, including robotics, open up the possibility of expanding production of even the most labor-intensive crops without hiring any low-skilled workers at all.

REFERENCES

Agerholm, H., 2018. UK crops let to rot after drop in EU farm workers in Britain after brexit referendum. February 5, In: Independent. http://www.independent.co.uk/news/uk/home-news/uk-crops-eu-farm-workers-brexit-referendum-rot-manpower-recruitment-numbers-a8194701.html.

Aquacraft Water Engineering and Management (for the California Department of Water Resources), 2011. California Single Family Water Use Efficiency Study. http://www.aquacraft.com/2015/07/28/california-single-family-water-use-efficiency-study/. [(Accessed 22 April 2016)].

Arax, M., Wartzman, R., 2005. The King of California: J.G. Boswell and the Making of a Secret American Empire. .

Central Europe Review. 1999 vol. 1(21), 15 November 1999; http://www.ce-review.org/99/21/vidali21.html

Chou, B., 2014. California Snowpack and the Drought. National Resources Defense Council. Fact Sheet 14–4-A, March 31, 2014, https://www.nrdc.org/resources/california-snowpack-and-drought. [(Accessed 22 April 2016)].

COA: Jobs to Workers, 2006. Rural Migration News. vol. 12(3)July 2006. https://migration.ucdavis.edu/rmn/more.php?id=1132.

Commission on Agricultural Workers, 1993. Report of the Commission on Agricultural Workers. Washington DC. https://catalog.hathitrust.org/Record/008306631.

Craig, R.B., 1971. The Bracer Program: Interest Groups and Foreign Policy. University of Texas Press, Austin.

Filipski, M., Taylor, J.E., 2009. Do the effects of free trade vary by gender? CAFTA and the rural Dominican Republic. Work. Pap., Dept. Agric. Econ., Univ. Calif., Davis.

Fogel, R.W., Engerman, S.L., 1980. Explaining the relative efficiency of slave agriculture in the antebellum south: reply. Am. Econ. Rev. 70 (4), 672–690.

Fouratt, C.E., 2016. Temporary measures: The production of illegality in Costa Rican immigration law. PoLAR 39 (1), 144–160.

Fuller, V., 1940. The supply of agricultural labor as a factor in the evolution of farm organization in California. Unpublished PhD Diss., Univ. Calif., Berkeley. Reprinted in LaFollette Comm., Violations of Free Speech and the Rights of Labor (19778-894). Senate Educ. Labor Comm., Washington, DC.

Gibson, J., McKenzie, D., 2014. Development through Seasonal Worker Programs: The Case of New Zealand's RSE Program. World Bank Policy Research Working Paper 6762.

Hakim, J., 1995. A History of Us: War, Peace and all that Jazz. Oxford University Press, New York.

International Organization for Migration, 2001. Binational Study: The State of Migration Flows between Nicaragua and Costa Rica. http://www.rcmvs.org/investigacion/BinationalStudyCR-Nic.pdf.

Johnson, K., 2011. Hiring Locally for Farm Work Is No Cure-All. In: The New York Times. October 5, 2011, https://www.nytimes.com/2011/10/05/us/farmers-strain-to-hire-american-workers-in-place-of-migrant-labor.html. [(Accessed 22 May 2018)].

Kitroeff, N., 2017. How this garlic farm went from a labor shortage to over 150 people on its applicant waitlist. In: Los Angeles Times.February 9, 2017http://www.latimes.com/business/la-fi-garlic-labor-shortage-20170207-story.html. [(Accessed 22 May 2018)].

Marosi, R., 2015. Labor Talks Over Mexican Farm Worker Wages Wobble. In: Los Angeles Times. March 26, 2015, http://www.latimes.com/local/california/la-me-baja-farmworkers-20150326-story.html.

Martin, P., 1996. Promises to Keep: Collective Bargaining in California Agriculture. Iowa State University Press, Ames, IA. 416 p.

Martin, P.L., 2003. Promise Unfulfilled: Unions, Immigration and the Farm Workers. Cornell University Press, Ithaca and London.

Martin, P.L., 2009. Importing Poverty? Immigration and the Changing Face of Rural America. Yale University Press, New Haven.

Martin, P.L., Taylor, J.E., Hardiman, P., 1988. California farm workers and the SAW legalization program. Calif. Agric. 42 (6), 4–6.

Martin, P.L., Midgley, E., Teitelbaum, M.S., 2002. Migration and development: whither the Dominican Republic and Haiti? Int. Migr. Rev. 36 (2), 570–592.

Martin, P.L., Fix, M., Taylor, J.E., 2006. Migrants in U.S. Agriculture. Chapter 2, In: The New Rural Poverty. The Urban Institute Press, Washington, DC.

Massey, D.S., Singer, A., 1995. New estimates of undocumented Mexican migration and the probability of apprehension. Demography 32 (2), 203–213.

Massey, D.S., Arango, J., Hugo, G., Kouaouci, A., Pellegrino, A., Taylor, J.E., 1993. Theories of international migration: a review and appraisal. Popul. Dev. Rev. 19 (3), 431–466.

McWilliams, C., 1939. Factories in the Field: The Story of Migratory Farm Labor in California. Little, Brown and Company, Boston.

Migration Advisory Committee, 2017. EEA-Workers in the UK Labour Market: A Briefing Note to Accompany the Call for Evidence. August 4, 2017.

Migration News, 2009. Africa: Migrants, SA, USAID, vol. 15(4), October, and Southern Europe, vol. 15(2), April.

Rural Migration News, 2004. Africa: Land, Cotton, vol. 10(1) January.

Rural Migration News, 2007. Canada, UK, Middle East, vol. 13(2) April

Rural Migration News, 2008. Spain: Strawberries, Migrants, vol. 14(2) April,

Rural Migration News, 2009. UK: Migrants, vol. 15(1) January.

Taylor, J.E., 2010. Agricultural labor and migration policy. Ann. Rev. Resour. Econ. 2, 369–393. http://www.annualreviews.org/https://doi.org/10.1146/annurev-resource-040709-135048.

Taylor, J.E., Thilmany, D., 1993. Worker turnover, farm labor contractors, and IRCA's impact on the California farm labor market. Am. J. Agric. Econ. 75 (2), 350–360.

Tzortzis, A., 2006. When Germans Join Migrant Field Hands, the Harvest Suffers. CSM. May 24, 2006, https://www.csmonitor.com/2006/0524/p01s04-woeu.html. [(Accessed 22 May 2018)].

U.S. Dept. Labor, 2005. Findings from the National Agricultural Workers Survey (NAWS) 2001–2002. A Demographic and Employment Profile of United States Farm Workers. U.S.

Dept. Labor, Off. Assist. Secr. Policy, Off. Program. Policy, Res. Rep. No. 9. March, http://www.doleta.gov/agworker/report9/toc.cfm.

U.S. Dept. Labor, 2017. Findings from the National Agricultural Workers Survey (NAWS) 2013–2014.

Yu, B., 2005. Immigration Multiplier: A New Method of Measuring the Immigration Process (Doctoral dissertation). Brown University.

FURTHER READING

Kurland, P.B., Lerner, R. (Eds.), 1987. The Founders' Constitution. University of Chicago Press, Chicago. http://press-pubs.uchicago.edu/founders/. [(Accessed 19 March 2016)]. Ch. 15, Doc. 32.

Taylor, J.E., Charlton, D., Yunez-Naude, A., 2012. The end of farm labor abundance. Appl. Econ. Perspect. Policy 34 (4), 587–598.

Chapter 7

Farm Labor Organizing From Cesar Chavez and the United Farm Workers to Fair Foods

It's ironic that those who till the soil, cultivate and harvest the fruits, vegetables, and other foods that fill your tables with abundance have nothing left for themselves.

Cesar Chavez

Human slavery is a not a thing of the past, but an ugly crime that still continues to afflict our communities.

A. Brian Albritton, US Attorney for the Middle District of Florida

Most farm workers around the world are poor and vulnerable, with little educa-tion and few alternatives to doing hired farm work. In the United States, Cesar Chavez and the United Farmworkers Union faced the challenge of organizing a workforce that was seasonal and mostly immigrant, without legal status. Farm labor advocacy is more daunting in an era of complex, global food supply chains. This chapter uses economic tools to analyze the successes and failures of farm labor organizing as well as new strategies that leverage the food supply chain in an effort to improve agricultural working conditions in rich and poor countries.

Imagine a country in which farmworkers are brutally beaten, paid below-poverty wages, kept in debt, and chained by their hands so that they cannot leave and look for better jobs. They are locked in box trucks, crates, and sheds.

You have just put yourself in the United States of America—specifically, Immokalee, Florida, in 2008.

This is not fiction or hearsay. In 2008, four Immokalee family members were sentenced in federal court for enslaving and brutalizing migrant tomato pickers from Mexico and Guatemala. Chief Assistant US Attorney Doug Mol-loy called it one of Southwest Florida's "biggest, ugliest slavery cases ever."

One of the ugliest. This was not an isolated case. Around the same time, 12 people were convicted of crimes involving enslaving farmworkers in Flor-ida. They were the ones who got caught.

The Farm Labor Problem. https://doi.org/10.1016/B978-0-12-816409-9.00007-0

155

This chapter presents a brief history of efforts to organize and improve the earnings and working conditions of hired agricultural workers. Most farm workers around the world are poor and vulnerable, with little education and few alternatives to doing hired farm work. In Chapter 3 we saw that a mover-stayer model sends workers to the sector that offers them the most benefits in terms of expected earnings, working conditions, and other amenities. Most hired farm workers are so constrained, however, that moving out of agriculture is almost not an option. In villages from Mexico to India and Malawi, hired agricultural workers typically are from households with little or no land, schooling, or other assets. On commercial farms in developing countries, like Mexico's agro-export state of Sinaloa, one often finds farmworkers who are migrants from poor zones within the same country—in Mexico's case, from the poor southern states of Chiapas and Oaxaca. In rich countries, most hired farmworkers are immigrants, many without the legal right to be there. Throughout history, farm worker advocates have had to face the realities of legal systems that do not support farm worker rights and often criminalize farm labor organizing. They also have faced the challenges of organizing a workforce that is difficult to organize because farm workers are vulnerable, poor, constantly on the move, and strive to remain invisible to the authorities. The success of Cesar Chavez and the United Farm Workers (UFW) established the State of California as one of the most committed protectors of farmworker rights in the world. But even in California, protecting farmworkers' livelihoods and basic rights along with interpreting complex legal codes continues to be a challenge.

THE PLIGHT OF FARM WORKERS IN AMERICA

Gaining access to immigrant farmworkers is the key to solving the farm labor problem from the farmers' perspective. As the US workforce transitioned away from hired farm work, labor-intensive agricultural production in the United States was able to expand because farmers could hire workers from Mexico. Working conditions were variable. However, with few exceptions, the work was seasonal and the wages were low. Individual farm workers, vulnerable and with limited options, had little influence to change their working conditions. To demand higher wages and better working conditions required organization. Organizing and representing a largely immigrant seasonal workforce is the great challenge confronting farm labor organizers.

When the Dust Bowl temporarily increased the supply of domestic workers to farms, writers, politicians, journalists, and photographers publicized the plight of America's farmworkers. There were efforts to improve the lives and working conditions of farmworkers by the American public and politicians, who were moved by images of people who looked like them, living in tent cities near the fields, buying food and necessities in the company stores which often charged exorbitant prices, and taking their children out into the fields to work alongside them. Follow-the-crop migration interrupted children's education. Nevertheless, farmworkers and farm labor organizers found themselves up against a powerful coalition of farmers, local politicians, and police willing to violently crack down on worker unrest and even bulldoze entire labor camps at short notice. These actions inspired John Steinbeck's classic novel *Grapes of Wrath*.

Braceros fared no better. A lack of Mexican government labor inspectors, especially on the more isolated farms of Oregon and Washington states, made enforcement of Bracero contracts and labor rules problematic. According to one observer, "In 1943, ten Mexican labor inspectors were assigned to ensure contract compliance throughout the United States; most were assigned to the Southwest and two were responsible for the northwestern area (Gamboa, 2000)." The Associated Farmers Incorporated of Washington State strove to keep farmworker pay down and prevent labor organizers from entering the fields, enlisting the assistance of private security forces, the state highway patrol, and even the National Guard. Closer to the border, in California and Arizona, employers threatened Braceros with deportation, knowing they could easily replace them with new Braceros.

The 1960 CBS documentary *Harvest of Shame*, hosted by Edward R. Murrow, interviewed farmworkers, their families, and their children's schoolteachers. Conditions for farmworkers and their families were harsh. Farmworkers managed to keep their families fed, but there was little left to invest in a more stable future. Children of farmworkers usually began school when they were young, but they rarely completed a school year because their families had to move to the next harvest. Most dropped out early because their families needed them to work.

CESAR CHAVEZ AND THE UNITED FARM WORKERS UNION

> *People were crowding into the building. Inside were a sea of Latinos. Some were UFW members. The rest were their friends, families and supporters. I saw a few Anglos like myself…The word spread that Cesar Chavez was soon to arrive. He would end his fast with the body and blood of Christ. Communion…Large red and black UFW banners, with the big black Aztec eagle image, hung around the hall…Joan Baez was introduced to the crowd and she made her way to the front of the room, guitar in hand…Everyone began to sing along with "We Shall Overcome." The audience joined hands and swayed back and forth as we sang…A small crowd of people entered the room from the left and proceeded to the front of the room now crowded with people standing. Cesar Chavez had arrived. Weakened by his fast, he rode in a wheelchair pushed by friends and family. The communion ritual began.[1]*

---

1. Eyewitness account of a UFW rally in Superstition Review: An Online Literary Magazine, Arizona State University, post by Jeff Falk; https://blog.superstitionreview.asu.edu/tag/macayos/.

Labor strikes, hunger strikes, marches, boycotts, celebrities on stage. These were the weapons of Cesar Chavez and the UFWs as they carried out the largest campaign ever in the United States, and probably the most celebrated in the world, to organize farmworkers and improve their salaries and working conditions. The laws of the land were not on farm labor organizers' side when, in 1962, Cesar Chavez, a Mexican-American farmworker from humble beginnings, cofounded the National Farmworkers Association, later to become the UFW, with Dolores Huerta. At that time, most nonfarm workers across America had the right to form or join unions and engage in strikes and other activities to improve working conditions—but not farm workers. The National Labor Relations Act (NLRA) of 1935 protected factory workers, but it explicitly excluded "domestic workers and farm workers."

Chavez joined Filipino-American farmworkers in a grape strike in Delano, California, on September 8, 1965, to protest years of poor pay and working conditions. He realized that a strike was not enough, and the union and its workers were vulnerable. He needed the support of politicians, the media, and the American public.

Six months later, he led a historic 340-mile march from Delano to Sacramento, the California state capital. The grape strike and march attracted national attention—so much that grape growers could not ignore or crush it. The next year, in March, the US Senate Committee on Labor and Public Welfare's Subcommittee on Migratory Labor traveled from Washington DC to California to conduct hearings on the strike. Robert F. Kennedy, a member of the subcommittee, openly supported the striking workers. These events sparked similar movements in other US states and led to the creation of the independent farmworker unions Obreros Unidos (United Workers) in Wisconsin (1966), the Farm Labor Organizing Committee (FLOC) in Ohio (1967) and the Texas Farm Workers Union (1975). The UFW's slogan, "*si se puede*" (yes, it can be done), echoed through the crowds.

A watershed movement supporting farmworker rights had begun. The stakes were high, but it was not easy, despite growing political and media support. The grape strike dragged on, lasting more than 5 years. Chavez and the UFW needed new weapons.

Beginning in 1968, Chavez subjected himself to "spiritual fasts" that garnered him and the UFW more national attention. Now on a national stage, he and the UFW went straight to American consumers, urging them to boycott nonunion head lettuce and table grapes during the harvest season. Consumer boycotts proved more effective than strikes at harvest time, for reasons that we see later on in this chapter. On August 23, 1970, the UFW launched the "Salad Bowl Strike," the largest farm worker strike in US history.

Never before had a farm worker organization wielded so much power, including the influence to force growers' hand by appealing to millions of consumers. The State of Arizona responded by making strikes and consumer boycotts illegal at harvest time. Growers tried to discourage their workers from joining unions—some, by offering improved wages and working conditions without union contracts. In the lemon groves of Ventura, California, the Coastal Growers Association (CGA) offered farmworkers more stable employment by synchronizing their activities across member farms. By doing this, it reduced its number of pickers from 8,517 in 1965 to 1,292 in 1978, while paying an average hourly wage ($5.63) that was more than twice the minimum wage at the time ($2.65) and offering benefits that included health insurance, paid vacations, and subsidized housing (Mamer and Rosedale, 1975).

Chavez' aggressive but nonviolent organizing tactics forced growers in California and Arizona to recognize the UFW as the union representative of 50,000 farmworkers. Chavez and the UFW achieved their goal of extending rights long enjoyed by factory workers to the fields of California. On June 4, 1975, Governor Brown signed into law the California Agricultural Labor Relations Act (ALRA), still regarded as a landmark in US labor law. The ALRA for the first time established the right to collective bargaining for agricultural workers. It encouraged and protected:

> The right of agricultural employees to full freedom of association, self-organization, and designation of representatives of their own choosing, to negotiate the terms and conditions of their employment, and to be free from the interference, restraint, or coercion of employers of labor, or their agents, in the designation of such representatives or in self-organization or in other concerted activities for the purpose of collective bargaining or other mutual aid or protection.

The ALRA established the California Agricultural Labor Relations Board (ALRB) to ensure that the ALRA would be carried out as intended.

THE DECLINE OF FARM LABOR UNIONS

A half century after the grape strike and march on Sacramento, almost nobody in California's farm workforce is a member of the UFW or any other farm-worker union. By 2006, the United Farmworkers Union had fewer than 25 con-tracts (Miriam Pawel, 2006). During their heyday, Chavez and the UFW were torn between representing workers and changing the world, between being a union or a political movement. Chavez realized that he would have to change the law of the land in order to give farmworkers the right to organize and improve their circumstances. He managed to change the law of one land: California, which employs far more seasonal farm workers than any other state. Despite this, few other US states followed suit, and the cards were stacked against the UFW's efforts to unionize the farm workforce in California or elsewhere. Many studies have tried to explain why, but from an economic point of view there seem to be three main explanations for the failure of farm labor unions:

First, farmworkers are notoriously difficult and costly to organize. It is chal-lenging enough to organize a stable factory workforce of legal workers. The traits that make agriculture different also make agricultural workers hard to organize. Seasonal agricultural workers move from farm to farm. The ALRB had to struggle with the question of what constituted a farm's workforce. Which workers could vote to join a union—the ones present on the farm at harvest time, or the smaller number of year-round workers? With seasonality of labor demand, the date at which the union election takes place might make all the difference in determining whether a farm's workers voted to unionize.

Second, many farmworkers failed to embrace the idea of joining a union. To understand why, we can use a mover-stayer model to predict when a farm worker is likely to join a union and when he is not. A worker will join a union if the expected gain in income and working conditions from doing so exceeds the costs. In the 1970s, most California farmworkers were immigrants from Mexico, and many or most did not have legal permission to work in the United States. How could the UFW convince these individuals that their lives would be better in the union? That it would be worth paying union dues, and then hope that the union would strike a good contract with the farmers who employed them? That standing up to be counted in a union election would not expose them to immigration authorities intent on deporting them to Mexico, or prevent them from being employed by other farmers in the future? The California ALRB tried to keep these abuses from happening, but its powers were limited, particularly when it came to the federal government's efforts to deport illegal immigrants. Many farmworkers realized that the best way to move up the US job ladder was to leave agriculture altogether, not to stick it out in farm work and join a union.

Third, the UFW struggled to organize the farm workforce during an era of farm labor abundance. The end of the Bracero Program in 1964 (Chapter 6) created a brief window in which the farm labor supply from Mexico switched from being legal (at least in part) to undocumented. This was a moment in history when farmers worried about having access to the workers they needed and unions could flourish. However, with hundreds of thousands of young rural Mexicans willing to migrate illegally to work on US farms, it did not take long for farm labor surpluses to emerge, as the supply of farmworkers from Mexico exceeded the demand.

The surge in unauthorized immigration that followed the Bracero Program was not popular among labor organizers. The UFW recognized the difficulty of organizing a farm workforce in which more and more workers were illegal and afraid of deportation. To add to this challenge, the seasonality of farm labor demand caused the farm workforce to vary hugely during the year. Workers moved to more stable and higher paying nonfarm jobs as soon as they were able; construction and other sectors competed for low-skilled farmworkers. New unauthorized workers came into the United States to fill the void this left in the fields. Thus, the farm workforce was seasonal and ever changing, and this complicated organizing efforts enormously.

In the hope of discouraging immigration, the UFW called for laws making it illegal for farmers to hire unauthorized immigrants. It found itself in the awkward position of representing an increasingly illegal workforce, while at the same time joining conservative politicians in supporting legislative action to crack down on illegal immigration and the employment of unauthorized immigrants.

Remarkably, it even cooperated with the US Immigration and Naturalization Service (INS) to apprehend some illegal immigrants and send them back to Mexico. The UFW recanted this position later on, letting go of its anti-immigration stance and supporting amnesty provisions in the 1986 IRCA. Nevertheless, its inconsistent and contradictory positions on immigration illustrate the challenges of organizing a constantly changing, seasonal immigrant farm workforce. As the farm workforce became more illegal, fed by an elastic supply of new workers from rural Mexico, unions saw their hold on farm organizing slip away.

Cesar Chavez' failure to unionize the farm workforce does not diminish the importance of the political victories that he and the UFW achieved. Advocates' efforts to improve farmworkers' wages and working conditions during this period offer insights into the most effective ways to support farmworkers in places where the law of the land is not on their side.

HOW THE "SALAD BOWL STRIKE" BACKFIRED

One of the most poignant events illustrating the difficulties of organizing farm labor is the lettuce workers strike in 1970. On July 17, 1970, 6,000 drivers and packing workers went on strike in Salinas, California—the lettuce capital of the world. The strike, sponsored by the Teamsters, kept most of the lettuce crop from reaching consumers. Lettuce farmers plowed under thousands of acres of good lettuce.

A strike at harvest time might appear like an effective strategy for farm labor organizers to win concessions, but there was a catch: the lettuce price. Lettuce is perishable. Almost overnight, the price of lettuce tripled. Individually, losing most of a crop reduces a farmer's profits, but when the entire industry is unable to harvest its crop, the market price rises, and producers can actually benefit from decreased sales. One study found that the so-called "salad-bowl strike" of 1970 increased farmers' total profits![2]

Figs. 7.1 and 7.2 illustrate how a strike at harvest time can backfire on labor organizers. Fig. 7.1 shows an upward-sloping supply curve and downward-sloping demand for lettuce. Before the strike, the intersection of the supply curve, S_0, and the demand curve, D_0, gives a competitive equilibrium lettuce price of P_0 and quantity Q_0.

In Fig. 7.2, the strike reduces the quantity that reaches the market to Q_1, making the supply curve vertical at the point where the strike takes hold. The new supply is shown in bold as S_1. However, given the slope of the lettuce demand curve, the reduced supply of lettuce drives up the equilibrium lettuce price, from P_0 to P_1.

What happens to farm profit? Profit, also called the producer surplus, is the difference between total revenue and total cost. Total revenue before the strike is given by P_0Q_0, or rectangular area $(a+b)$ in Fig. 7.1. At each level of output,

---

2. This example is inspired by Carter et al. (1981).
the cost of producing an additional unit, or marginal cost, is given by the supply

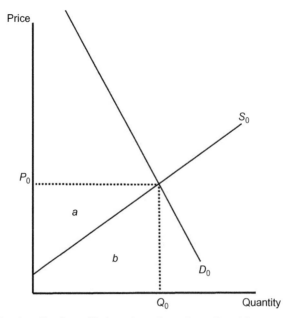

FIG. 7.1 Before the strike, the equilibrium price and quantity are P_0 and Q_0, respectively. Profit or producer surplus is represented by area a.

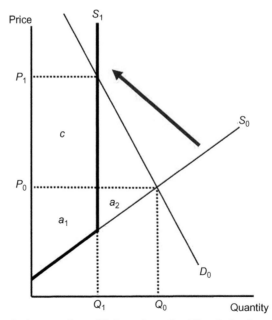

FIG. 7.2 The strike increases the equilibrium price to P_1 while reducing the quantity to Q_1. The change in farm profit is area $(a_1+c)-(a_1+a_2)=c-a_2$.

curve. The triangular area b is the total cost, or sum (in calculus, the integral) of marginal costs across all output levels from zero to Q_0. The prestrike profit, then, is represented by area a in Fig. 7.1.

After the strike, profit is calculated with the new price, P_1. The new profit is $(a_1 + c)$. The change in profit, then, is $(a_1 + c) - (a_1 + a_2) = c - a_2$. In this example, area c clearly is larger than area a_2, so profit increases.

The strike does create costs for producers. Farmers lose the area a_2, which they received before the strike, because of the reduced quantity of lettuce they sell in the market. But they gain area c. In this case, the higher price more than compensates for the quantity loss.

This does not necessarily mean that *all* farmers gain. Some might not be able to get as much of their lettuce to market as others. There are winners and losers, even if on balance the effect of the strike on profits is positive.

BOYCOTTS: A MORE EFFECTIVE BARGAINING TOOL

Beginning in 1965, Cesar Chavez and the UFW led a consumer boycott against table grapes. It became known as the Delano grape boycott, since it was launched in Delano, a town in California's rich Central Valley. The UFW relied on nonviolent grassroots tactics, with community organizing, marches, and calls to consumers not to buy table grapes. It gained national attention, and it had the effect that the farmworkers desired. In 1970, the UFW reached a collective bargaining agreement benefiting more than 10,000 table grape workers.

Why did the grape boycott work so well while the lettuce strike did not? We can gain insights from a simple economic model. The boycott convinced many consumers not to buy table grapes in order to support farm workers. The reduction in demand is shown by a shift in the demand curve from D_0 to D_1 in Fig. 7.3. Before the boycott, the aggregate producer profit was the sum of areas a_1, a_2, and d. Afterwards it was only area d. Fig. 7.3 shows that effective boycotts unambiguously reduce profits, because they lower *both* the quantity *and* the price. One large California grower exclaimed: "The grape boycott scared the heck out of the farmers, all of us" (Ferriss and Sandoval, 1998).

FARM LABOR UNIONS AROUND THE WORLD

Farm labor organizing is by no means unique to the United States. Many of the challenges facing farm labor unionization in the United States are shared throughout the world. The National Union of Agricultural Workers represented farm workers in the United Kingdom from 1906 to 1982; in 1982 it became the Agricultural Section of the Transport and General Workers' Union. An analysis of the union in 1964 indicated that the union had difficulties in recruiting members, and many joined because they had an accident, were sick, or were facing an eviction and required union services (Mills, 1964). The union had 3,500 branches, which made administrative organization difficult.

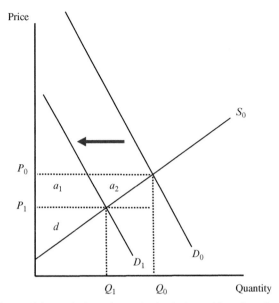

FIG. 7.3   The boycott's impact is shown by a reduction in demand from D_0 to D_1. This lowers the price to P_1. The profit loss is $(a_1 + a_2)$.

The European Federation of Food, Agriculture, and Tourism Trade Unions (EFFATs) currently represent 120 national trade unions, including agriculture, from 35 European countries. EFFAT maintains a voice in European policy to protect agricultural employment, secure fair pay in rural areas, and further social dialogue to improve working conditions (http://www.effat.org). Current EFFAT concerns focus, in part, on the European Union's Common Agricultural Policy (CAP). Many farms in Europe would not be viable without support and subsidies from CAP, so farm workers, like farmers, depend on CAP. In 2018, EFFAT called for an end to subsidies to employers that exploit workers or do not provide adequate working conditions, for written confirmation of employee contracts by the first day of employment, and for the European Union Framework directive on health and safety at work to be the criteria for distributing CAP direct payments. EFFAT contends that it makes these requests on behalf of 10 million agricultural workers.

THE FAIR FOOD MOVEMENT

After Florida's farmworker slavery cases in 2008, farm worker advocates realized that something had to be done, but they did not have the law on their side. There was no equivalent of California's Agricultural Labor Relations Act (ALRA), guaranteeing collective bargaining for farmworkers in Florida. The prospects of achieving a legislative solution to farmworker exploitation, like Cesar Chavez and the UFW did in California, were dim.

It was time to become an economist.

The Coalition of Immokalee Workers (CIWs) asked itself a simple question: Who benefits from exploiting Florida tomato workers? The answer took them from Immokalee, Florida to Irvine, California.

In 2001, the CIW picketed Taco Bell's (TBs) headquarters in Irvine, with mottos like "Taco Bell produces exploitation food" and "sweatshop tacos" (Munoz, 2003).

Taco Bell does not employ farmworkers. It demands a lot of tomatoes, though. The CIW held Taco Bell accountable for the treatment of farmworkers in its tomato supply chain.

The Coalition's argument went like this: Like other big buyers, Taco Bell leverages its market power to obtain low prices from its suppliers. This price pressure reverberates down the supply chain, as sellers squeezed from above put pressure on their suppliers to sell at rock-bottom prices. Farmworkers are the lowest link in the tomato supply chain. They have no one to pass the price pressure on to.

The protest immediately hit the *Los Angeles Times* and other major newspapers. It also got plenty of attention from Taco Bell, which like other corporations is sensitive to public opinion, particularly in an era when bad publicity can spread like wildfire on the internet and impact corporate profits.

The company's initial reaction to the protest seemed to be one of confusion. A Taco Bell spokesman said: "The coalition efforts are misdirected. They do not work for Taco Bell."

Four years later, on March 8, 2005, Taco Bell's parent company, Yum! Brands, Inc., agreed to all of the CIW's demands. These demands included payments by Taco Bell to farmworkers; an enforceable "Code of Conduct," which all Florida farmers supplying tomatoes to Taco Bell must adhere to; employers' respect for workers' human rights, even when those rights are not guaranteed by law; and complete transparency for all of Taco Bell's tomato purchases in Florida.

In 2010 the CIW launched the Fair Foods Program (FFP). In addition to measures like the ones in the Taco Bell agreement, the FFP includes worker-to-worker education sessions on the new labor standards set forth in the program's Fair Food Code of Conduct. It establishes a third-party monitor, the Fair Food Standards Council, which makes sure that farmers comply with the FFP's provisions and carries out regular audits and ongoing complaint investigation and resolution. Taco Bell and other participating buyers of Florida tomatoes pay a small Fair Food premium—"a penny a pound"—which tomato growers must pass on to workers as a line-item bonus on their regular paychecks. Nearly $15 million was paid into this fund between January 2011 and May 2014, substantially increasing Immokalee farmworkers' earnings (CIW Website, 2014a).

By 2014, the CIW had reached similar agreements with McDonalds, Burger King, Whole Foods, Subway, Bon Appétit Management Company, Trader Joe's, Chipotle Mexican Grill, Walmart, and others. These agreements are legally binding. They guarantee market consequences for any firm that does not comply—not to mention a lot of bad publicity. The participating companies commit to buy Florida tomatoes only from growers who are in good standing with the FFP. They must stop purchases from growers who fail to comply with the code of conduct.

Not surprisingly, very few growers fail to comply, because if you grow a lot of tomatoes, the alternative to selling to big players like Walmart and Taco Bell is not very promising.

As of 2015, the CIW's Fair Food Code of Conduct was implemented on 90% of Florida's tomato farms, covering approximately 30,000 acres and employing tens of thousands of workers.[3] Former US President Bill Clinton called the FFP "the most astonishing thing politically happening in the world we're living in today" (CIW Website, 2014b).

THE ECONOMICS OF FAIR FOOD: TURNING MARKET POWER ON ITS HEAD

The FFP takes the supply chain pressures that contributed to farmworker exploitation in the past and turns them on their head. The boycott targets the top of the market chain, but its biggest impact is at the bottom of the chain: on farmworkers.

Market power by big buyers sensitive to public opinion is the key to the FFP's success. The FFP strategy is less feasible if buyers do not have market power. The FFP would be difficult—and very costly—to implement in a world where many competitive buyers compete with one another. Visualize hundreds of importers across the globe willing to buy tomatoes from Florida farmers. Having a contract with Taco Bell would not be so important to farmers if they could sell their tomatoes to someone else at low cost without transferring profits to farmworkers. What makes the FFP work is that the CIW only has to target a few strategically chosen "big players" at the top of the tomato market chain.

Why has the Fair Food Program been so effective? Let us do the economic analysis.

We begin with the market for Taco Bell (TB) tacos, depicted in Fig. 7.4. The supply of tacos is given by Taco Bell's marginal cost curve. The demand curve reflects consumers' willingness to pay for Taco Bell tacos. The intersection of the two gives the equilibrium price and quantity of Taco Bell tacos.

---

3. You can learn more about the CIW Code of Conduct at http://www.fairfoodstandards.org/code. html.

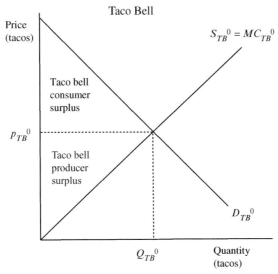

FIG. 7.4 The equilibrium price of TB tacos is p_{TB}^0, and the equilibrium quantity is Q_{TB}^0. The producer surplus, or profit, is given by the triangle between the horizontal price line and marginal cost curve.

This is clearly a simplification. Taco Bell produces more than tacos—burritos, for instance. In addition, it competes with a few other fast-food corporations that sell tacos (and burritos). In that case, it may appear that we should model the market for all tacos (and burritos), not just Taco Bell's. On the other hand, there is no question that, in this market space, the market for cheap fast-food tacos, Taco Bell is dominant.

Since Taco Bell is the dominant seller of tacos and a major consumer of tomatoes, we present a model that generalizes to the situation of a monopolist taco firm when consumers care about farm workers. The simplifications in this model allow us to explain why the Fair Food Program works by focusing on Taco Bell and on tacos.

In this figure, the TB market equilibrium is at price p_{TB}^0 and quantity Q_{TB}^0. The producer surplus, or TB profit, is the triangle between the price line and marginal cost curve.

As TB's taco production increases, so does TB's demand for tomatoes. Just as we simplified TB as having a single market, we also simplify the Florida tomato producers as having a single buyer—Taco Bell—whose recipe demands a certain amount of tomatoes per taco. Let us represent this amount of tomatoes per taco by the Greek letter γ ("gamma"). The demand for Florida tomatoes, then, is simply γ times Q_{TB}^0, as shown in Fig. 7.5. It is what economists call a "derived demand." It depends on the equilibrium quantity of TB tacos. Florida tomato farmers' supply of tomatoes is represented by their marginal cost curve in Fig. 7.5.

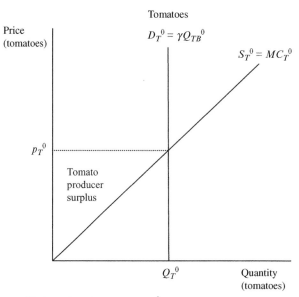

Tomatoes

Price (tomatoes)

$D_T^0 = \gamma Q_{TB}^0$

$S_T^0 = MC_T^0$

p_T^0

Tomato producer surplus

Q_T^0

Quantity (tomatoes)

FIG. 7.5 The equilibrium price of tomatoes is p_T^0, and the equilibrium quantity is $Q_T^0 = \gamma Q_{TB}^0$. The tomato farmers' producer surplus, or profit, is given by the triangle between the horizontal price line and marginal cost curve.

From Figs. 7.4 and 7.5, we can see that the Taco Bell and Florida tomato markets are linked together by TB's demand for Florida tomatoes. Anything that impacts TB also impacts Florida tomato growers, through the derived demand for tomatoes.

Now consider the effects of a successful consumer boycott of Taco Bell. Cesar Chavez and the UFW taught the CIW that boycotts can be an effective way to force employers to meet worker demands for improved wages and working conditions. The UFW carried out boycotts of specific commodities, particularly table grapes and head lettuce. The CIW took the boycott to a new level, by targeting a multinational corporation that bought tomatoes from growers who exploited farm workers: Taco Bell.

The CIW's goal was to convince consumers that TB tacos leave a bad taste in their mouth, because TB buys tomatoes from farmers who brutally exploit farm workers. If CIW can pull off a consumer boycott of Taco Bell, what will this do to the market equilibria in Figs. 7.4 and 7.5?

Fig. 7.6 shows the impact of a successful boycott on the market for TB tacos. The demand curve shifts inward: at the same taco price, consumers demand fewer tacos than before the boycott. This results in a new TB market equilibrium, with a lower taco price (p_{TB}^1) and lower quantity of tacos (Q_{TB}^1). You can see in Fig. 7.6 that the TB producer surplus not only drops—it decreases disproportionately with the fall in TB price, by an amount equal to areas

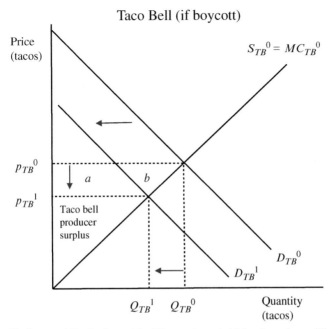

FIG. 7.6 The boycott shifts the demand for TB tacos inward, driving down the equilibrium price and quantity of TB tacos. TB's producer surplus, or profit, falls by an amount equal to areas $a+b$ in the figure.

$a+b$ in the figure. That is because the top part of the producer surplus triangle is much wider than the bottom part.

The contraction in the TB market spills over into the Florida tomato market, because TB's demand for tomatoes decreases. Fig. 7.7 shows the new Florida tomato market equilibrium. The quantity of tomatoes falls by γ times the drop in TB demand for Florida tomatoes. This drives down the price of Florida tomatoes (to p_T^1). The producer surplus, or profit, of Florida tomato producers falls disproportionately, by an amount equal to areas $c+d$ in Fig. 7.7.

In short, the boycott, if successful, would be costly to Taco Bell as well as to Florida tomato farmers. How much would each one be willing to pay to avoid the boycott?

The most that Taco Bell would be willing to pay to avert a boycott is the loss in producer surplus that the boycott would create—that is, areas $a+b$ in Fig. 7.6. The most that Florida tomato growers would be willing to pay is the sum of areas $c+d$ in Fig. 7.7.

These changes in producer surplus are the bargaining space that CIW can work with in its negotiations with TB and Florida tomato farmers. CIW presents its demands to TB. If TB refuses to meet those demands, CIW calls a boycott of

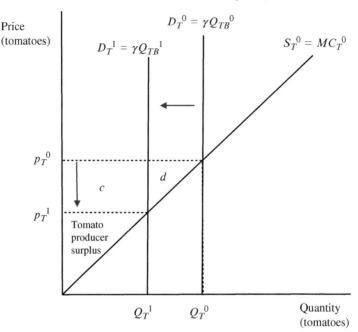

FIG. 7.7 The boycott decreases the derived demand for tomatoes, driving down the tomato price. Tomato growers' producer surplus, or profit, falls by an amount equal to areas $c + d$.

TB. In this age of the internet, when things go viral in a hurry, consumers around the country learn real time that TB indirectly exploits farmworkers, through its demand for Florida tomatoes. If consumers care about the welfare of workers who produce the food they eat, TB's demand shifts inward.

TB has two options. One option is to hold firm and not give in to CIW's demands. In this case, its profits fall by an amount equal to areas $a + b$ in Fig. 7.6. Alternatively, it can sign an enforceable contract with CIW, costing an amount less than areas $a + b$. The costs of this contract include the "penny-a-pound" pledge, in which TB would pay an additional penny for every pound of tomatoes it purchases to go directly to the farmworkers, plus whatever other costs are involved in meeting CIW's demands for transparency in all tomato purchases from Florida farmers.

If Taco Bell chooses to enter into a Fair Foods agreement with CIW, there might be other benefits. It can tell consumers that it sells "Fair Food" tacos. The CIW will have an incentive to back Taco Bell up on this, endorsing its product as Fair Food compliant. This could boost consumers' demand for TB tacos.

The tomato farmers face similar options. They could refuse to sign on to the FFP's code of conduct and be unwilling to pass along the "penny-a-pound" fund to their farmworkers. In that case, if TB has signed on to the FFP, the farmers will lose TB as a (very big) buyer. They could look for some other corporation to buy their tomatoes, while continuing to exploit their farm workers. But who could take TB's place as a buyer of their tomatoes? As other big buyers sign on to the FFP—Wendy's, McDonald's, Burger King, WalMart—the tomato farmers who refuse to go along with the program are left with few options. Not only that, they leave themselves open to court cases alleging "slavery in the fields."

Alternatively, farmers can sign an enforceable contract with the CIW, which costs them less than the sum of areas c and d in Fig. 7.7. They can also tell all their clients that they support Fair Food. This is likely to increase the demand for Florida tomatoes by other big firms, who might be willing to pay more for "Fair Food" tomatoes.

FAIR FOODS ON A COUNTRY SCALE?

Most of the fruit, vegetable, and horticultural production on the large irrigated farms in northwestern Mexico is exported to the United States. Countries, like companies, can be sensitive to how the world perceives them. On top of this, a few buyers account for a large part of Mexico's agricultural export demand, including some, like Walmart, that are already part of the CIW's Florida tomato agreement.

In December 2014, the *Los Angeles Times* ran a series of articles called *Product of Mexico*, which documented abuses of workers on fruit and vegetable farms in Mexico. It uncovered blatant abuses of labor laws and disregard for human rights, including people living under plastic tarps in labor camps, child labor, and de-facto slavery in the fields: Farmers essentially held workers captive by withholding their wages until after the harvest was over and they did not need them anymore. Most of the workers were members of indigenous groups who had little schooling, often did not speak Spanish, and had few outside options.

It seemed like *deja-vu*: extreme exploitation of farmworkers in the fields, but in Mexico instead of Florida.

On March 6, 2015, Mexican authorities launched a highly publicized crackdown on farms that allegedly violated Mexican labor laws and farmworkers' basic human rights. On May 14, the Mexican government, in an unprecedented move, pledged to subsidize farmworker wages.

Is there a connection between the *Los Angeles Times* series and the Mexican government's crackdown on exploitative farmers? *The Times* wrote that "Some experts said the crackdown was probably ordered in response to heightened awareness of farm labor abuses, such as those described in *The Times*' series, in Mexican media reports and by human-rights activists." A Mexican researcher was quoted as saying: "Officials seem to be more open to addressing the problem (Marosi, 2015)."

If both the US food industry and the Mexican government are sensitive to public opinion, this can work in favor of FFP-like efforts to protect farmworkers and improve their welfare.

"PRICING IN" THE ETHICAL TREATMENT OF FARMWORKERS

Just as consumers are willing to pay a premium for the quality of their produce, might they be willing to pay a premium for the quality of agricultural production practices? Consumers' willingness to pay for organic produce is one example. Many consumers buy organic, but it is more costly to do so. Growing organic crops generally is more expensive than using chemical-intensive methods.

Why are so many consumers willing to pay the price? One reason is health. Pesticide-free fruits and vegetables are likely to be more healthful to consume over one's lifetime. Another reason is a concern for the environment. Organic farming means fewer synthetic toxic chemicals on the land and in the water table, and it promotes soil conservation. Consumers' willingness to pay for these things is what makes the organic produce market work.

Are consumers willing to pay for ethical treatment of farmworkers, too?

A lot of passions fly around farm worker abuse. Some agricultural producers undoubtedly favor treating farmworkers well on ethical grounds. Like organic farming, though, this costs money. Unless there is a compelling economic incentive for *all* farmers to treat farm workers well, some farmers and farm labor contractors will gain at the expense of others. They will pay dismal wages, knowingly violate labor laws, or simply turn a blind eye to abuse in the fields. Unscrupulous farm employers will have a competitive advantage in the marketplace, because it obviously costs money to improve wages and benefits and monitor supervisors to make sure they adhere to ethical and legal labor standards. That is, unless consumers can see which growers provide adequate working conditions and are willing to pay a higher price to purchase food grown on "fair labor" farms.

Can the food market price in ethics? The CIW approach attempts to make this happen by creating negative consequences for big buyers who benefit from worker exploitation at the bottom of the supply chain. It forces these buyers to "price in" the ethical treatment of farmworkers.

Campaigns to raise the public's awareness of farmworker exploitation represent an effort to make consumers price in ethics, much like they do with organics. Ethics might be a harder sell than organics, though. The 1960 documentary *Harvest of Shame* brought the plight of America's farm workers into the public view.[4] Edward R. Murrow, who narrated the documentary,

---

4. You can see this documentary at: http://ciw-online.org/blog/2014/11/food-chains-harvest/.

explained: "We present this report on Thanksgiving because, were it not for the labor of the people you are going to meet, you might not starve, but your table would not be laden with the luxuries that we have all come to regard as essentials."[5]

Some observers praise *Harvest of Shame* for strengthening the push for new legislation, like funding for health services to migrant workers and education for migrants' children. It might have influenced public opinion in ways that made Cesar Chavez' and the UFW's boycott campaigns a little bit easier, but it did not result in a widespread consumer demand for ethical treatment of farmworkers. Most consumers, it seems, either are ignorant or else try not to think about unethical labor practices, even when they are an ingredient of some of the foods they consume.

USING LABELS TO HELP CONSUMERS MAKE MORE INFORMED DECISIONS

CIW's initial success was to engage major food corporations like Taco Bell in the fight against unfair labor practices in the field. Since Taco Bell and other fast food corporations represent a large share of Florida's tomato buyers, this strategy was highly effective. However, CIW's long-term vision extends beyond tomato farms in Florida. Advancement in the fair foods movement requires getting the support of consumers, as well.

In 2014, CIW released a "Fair Foods" label in several major grocery stores to distinguish tomatoes grown on farms that comply with good labor practices. To the consumer, each tomato in the store looks and feels the same, regardless of whether the workers who grew and harvested the tomato were dispossessed of their rights. However, some consumers are willing to pay a premium to guarantee that the farm workers who produce their food were paid adequately and treated well. The "Fair Foods" label enables consumers to distinguish tomatoes grown by farmers who comply with fair labor standards from tomatoes grown by farmers who do not. The label signals that the farm where the tomato was grown meets the standards set by the CIW for fair labor practices.

Food labels have been used for years to inform consumers and empower them to demand the kinds of products and qualities they desire, even if they cannot observe certain qualities directly. For example, organic food labels empower consumers to demand environmental and quality traits that they value in the produce they consume. The FDA uses health labels to inform consumers about certain nutritional qualities of food products.

The "Fair Foods" label may be an effective strategy to enforce more ethical labor practices, but it will only be effective if certain criteria are met. The "Fair Foods" label will only be effective if it meets the following three requirements:

---

5. National Public Radio, http://www.npr.org/2014/05/31/317364146/in-confronting-poverty-harvest-of-shame-reaped-praise-and-criticism.

First, it must be something that consumers recognize and trust, with easily identifiable criteria for producer compliance. If too many labels flood the market with different certification standards, product differentiation becomes muddled. The differences between labels with similar objectives but different certification criteria become confusing. This defeats the objective of the label, which is to help consumers make better-informed decisions.

Second, compliance with the label's labor standards must be effectively monitored and enforced. If fair labor practices are not strictly enforced as part of the certification process, then the label loses its credence.

Third, consumers must be informed. Many consumers care about agricultural labor practices and are willing to pay a premium to assure that their food was grown using fair labor practices. However, many are not aware of the fact that the food they consume may be grown using abusive labor practices.

Some large regional grocery stores, like Giant, and national grocery chains, like Whole Foods, have introduced the "Fair Foods" label. If consumers show high preference for "Fair Foods" certified tomatoes, the label is likely to appear in more grocery stores throughout the country and on more kinds of produce. Many consumers prefer buying produce that they are assured was grown using good labor practices. However, the success of the "Fair Foods" movement beyond tomato fields in Florida depends on providing consumers with the means to take responsibility for their own purchases.

COORDINATION OF CONSUMERS, UNIONS, AND NGOs ACROSS INTERNATIONAL BORDERS

The Fair Foods movement, beginning in Immokalee, Florida, is not the only movement of its kind. Worldwide, we can find examples of consumer preferences influencing the fair treatment of workers. These examples reinforce the hope that workers' rights will become better recognized throughout all stages of the supply chain regardless of how far an agricultural commodity might travel or how many stages of processing it might undergo.

One of the challenges in a global economy is the globalization of producer networks and growth of multinational enterprises. Procurement, distribution, and retailing are concentrated with the primary objective of meeting consumers' demands. This generally means providing the highest quality product at the lowest price. The rapid spread of supermarkets around the globe in the late 20th century placed unilateral buying power in the hands of relatively few mass distributers (Reardon et al., 2003). Supermarkets demand that agricultural producers supply goods at low prices and at the exact times and places needed. To meet demands for less expensive goods with more flexible contracts, producers further up the agricultural supply chain often cut wages, benefits, and job security for farm workers.

Labor contracting has grown in recent years, in large part because contractors can coordinate the supply of seasonal workers on short notice, match skills to appropriate tasks, and improve efficiency in the labor market (Barriento, 2013). In an economy where products are demanded in and out of season through global trade, coordination becomes critical. A case study of horticultural producers in the United Kingdom in 2004–05 interviewed 21,603 enterprises, of which 6,594 sold directly to supermarkets. 26% of the temporary workers employed through these enterprises were recruited directly by their employer and 54% were recruited through a labor contractor. Among the enterprises that did not sell directly to a supermarket, 74% of temporary workers were recruited directly by their employer and 46% were recruited through a labor contractor (Barrientos, 2013).

Labor contractors provide a critical service in the agricultural labor market, but the wedge contractors create between employers and workers creates potential space for exploitation of workers. Opportunities for exploitation increase the more contractors are wedged between employer and employee and the more dependent the worker is on the contractor. Foreign workers and particularly unauthorized workers are the most vulnerable.

Examples of efforts to reduce the potential for labor abuses include global civil society campaigns, reliance on trade unions and nongovernmental organizations (NGOs) to draw attention to abuses, and private sector "voluntary initiatives" that outline a code of conduct that corporations sign onto voluntarily. One of the weaknesses of voluntary initiatives is that they cannot monitor and guarantee compliance throughout the supply chain. This is of growing concern as global supply chains get longer and contractors further remove farm workers from the corporations at the end of the supply chain.

Rich-Country Legislation to Protect Poor-Country Workers

One solution to this problem, implemented in the United Kingdom, is a multistakeholder group called the Temporary Labor Working Group (TLWG), founded in 2002. The TLWG creates minimum standards for labor contractors. It is composed of trade unions, NGOs, and supermarket corporations. In 2004, it convinced the Parliament of the United Kingdom to pass the Gangmasters (Licensing) Act, which requires all labor contractors in the United Kingdom to register with the Gangmasters Licensing Authority and submit to monitoring. It also led to passage of the Basic Conditions of Employment Act, which designates "joint and several liability" holding producers accountable if contractors do not adhere to labor legislation, even if those contractors are located in another country. Since 2007, United Kingdom legislation has further required large and medium-sized companies to report on employee, social, and community issues within their value chains located overseas. These legislative actions help keep corporations and producers at the top of the value chain accountable for practices all along the chain, even when the chain includes many stages of processing or contracting that span international borders.

Exploiting Global Supply Chains

Market power concentrated in supermarkets near the consumer base may generate incentives and space for exploitative practices further up the supply chain. However, it also opens up the potential for consumers to demand improved working conditions, as evidenced by the FFP and other examples throughout the world.

Bananas and IFAs

One such example was a coordinated agreement between the banana giant Chiquita and the Latin-American Coordination of Banana Workers Unions (COLSIBA), signed in 2001. The remarkable, and encouraging, aspect of this agreement was that it coordinated workers' unions across international borders using International Framework Agreements (IFAs). These are agreements for minimum labor standards negotiated between Global Union Federations and Multinational Enterprises like Chiquita. IFAs are still relatively rare, but they grew in number from 11 in 2001 to 107 in 2015 (Colenbrander, 2016). As global trade gains strength, farm workers around the world are forced to compete with one another. A victory for workers' rights in one country is hardly a win if producers subsequently move their operation to the next country where workers are underpaid or otherwise mistreated. IFAs provide a potential solution to this dilemma by negotiating terms of employment across international borders.

The case of Chiquita and COLSIBA shows how unions, NGOs, consumer groups, and a multinational enterprise can work together to create a solution that benefits farm workers. COLSIBA began in 1993 and had members in seven countries (Ecuador, Colombia, Costa Rica, Panama, Nicaragua, Honduras, and Guatemala). This network of banana workers' unions encompassed 42 unions, approximately 42,000 workers, and the production of more than 78% of the world's banana exports in 2005 (Riisgaard, 2005).

By coordinating multiple smaller unions and exchanging pertinent information, COLSIBA could gain greater leverage and bargaining power than a single union alone. Coordinated effort launched a damaging international campaign in 1998 accusing Chiquita of violating international workers' rights on Chiquita-owned and supplier plantations. It accused Chiquita of political corruption, owning secret companies, poisoning workers by spraying crops with a duster while they were in the fields, and sabotaging trade union activity.

Chiquita agreed to meet with COLSIBA, and after a series of meetings it signed a framework agreement to protect workers' rights. The agreement affirmed the right of each worker to choose to belong to and be represented by an independent and democratic trade union and to bargain collectively. Chiquita also agreed to abide by the International Labor Organization (ILO) core conventions and Convention No. 135, which protects workers' representatives

(Riisgaard, 2005). The agreement provided for a review committee allowing unions to access the corporate leadership, bypassing local management that may be uncooperative with union leaders. This was the first IFA agreement reached among developing countries.

In practice, the agreement between Chiquita and COLSIBA had mixed results that varied regionally and by plantation owner. Case studies indicated that many workers and local managers were unaware of the rights outlined in the agreement, and interviews suggested that working conditions improved on Chiquita-owned plantations but stayed the same or even became worse on supplier plantations. Despite mixed results, this case study instills hope for future, improved agreements involving coordinated efforts of unions across multiple countries and multinational enterprises, even in markets where a few multinational enterprises enjoy an almost exclusive market share (Riisaard, 2005).

Chiquita was incentivized to comply with union demands for multiple reasons: (1) consumers were concerned about corporate practices, including workers' rights and environmental sustainability, (2) reports that Chiquita mistreated workers had gained a spotlight in the press, putting pressure on Chiquita to make public changes, and (3) reaching agreements with workers reduced the risk of strikes or other activities that might stall production or reduce efficiency.

Flower Power

Cooperation among multiple agencies is not always easy, often for political, cultural, or practical reasons. Cut flower production in Kenya is labor intensive. Exports to the European Union grew from $13 million USD in 1980 to nearly $300 million in 2007 (Riisgaard, 2008); they reached $423 million in the first 4 months of 2018 (Oxford Business Group, 2018). Seasonality of consumer demand for cut flowers, which peaks during select holidays, causes labor demand to be volatile, and the perishability of the product means that workers often work long hours to complete orders.

Growing consumer concern over working conditions in the cut flower industry prompted supermarkets, trade associations in importing and exporting countries, NGOs, and government bodies to place new standards and pressures on the flower industries that export to the European Union. Labor unions exist in Kenya's flower industry, but only 3,400 out of 50,000 workers were unionized in 2006 (Riisgaard, 2008). Labor NGOs are gaining presence in the Kenyan flower industry, but cooperation between unions and NGOs remains problematic. Unions claim that NGOs do not represent the workers, and NGOs claim that since few workers are unionized (and those that are unionized are almost exclusively male permanent workers), unions are unable to adequately represent all workers in the flower sector. To make matters more complicated, sometimes corporations are unwilling to work with unions or NGOs. This case study illustrates the complex relationships that can form between unions, NGOs, and private enterprises, sometimes to the detriment of the workers for whom they advocate.

The combined examples of the FFP initiated in Florida, the TLWG in the United Kingdom, and the success of an IFA between Chiquita and the Latin-American Coordination of Banana Workers Unions demonstrate that consumers can demand and ensure fair working conditions for farm workers despite long and complex global supply chains. The results are not perfect, but they illustrate that coordinated efforts can bring about changes and improvements along the global grocery supply chain, all the way to the rights of workers in the fields.

REASONS FOR HOPE

There are several reasons to be hopeful about the prospects for achieving better treatment of farmworkers.

First, assuming consumers do care, it is in the interest of big buyers who join the FFP and similar programs to make consumers aware of their commitment to the ethical treatment of farmworkers, just as it is in organic producers' interest to label their products as "organic." Labeling and advertising labor standards could influence consumers' willingness to pay for better treatment of farmworkers. For many people, having a clear conscience increases utility from food purchases.

Second, it is not necessary to change *all* consumers. Big impacts happen at the margin. Small changes in prices, and bad (or good) publicity that shifts consumer demand a little, often have disproportionately large influence on companies' purchasing decisions.

A third reason to be hopeful is that, as we shall see later in this book, the farm labor supply is declining. In rural areas of Mexico and many other farm labor exporting countries, fewer children are growing up to be farmworkers. In a labor-scarce world, employers must compete for a finite supply of workers, offering wages, benefits, and other amenities to attract good workers.

Farmworkers in northwestern Mexico come from the poorest parts of southern Mexico. In our labor supply model (Chapter 3), we saw that employers have to offer a wage at least equal to the reservation wage in order to induce workers to come to their farms. For an unskilled laborer in rural Oaxaca or Chiapas, that reservation wage is very low.

Over time, the rural transformation accompanying economic development expands schooling opportunities for children growing up in poor rural areas, creates incentives for women to have fewer children, and opens up nonfarm employment prospects. This raises opportunity costs and creates upward pressure on wages, benefits, and working conditions on commercial farms. Most importantly, labor scarcity strengthens farmworkers' bargaining position.

Growing farm labor scarcity potentially complements and reinforces farmworker advocacy efforts. It is likely to be easier to convince farmers to enter into a union contract or FFP agreement if they face pressure from a declining farm workforce. Smart farmers know that ethical and fair labor practices are a key to attracting and retaining workers in a changing and uncertain world, in which crops have to be harvested and workers have more options than they did in the past.

REFERENCES

Barrientos, S.W., 2013. 'Labour chains': analysing the role of labour contractors in global production networks. J. Dev. Stud. 49 (8), 1058–1071.

Carter, C.A., Hueth, D.L., Mamer, J.W., Schmitz, A., 1981. Labor strikes and the price of lettuce. West. J. Agric. Econ. 1–14.

CIW Website, 2014a. Worker-driven Social Responsibility (WSR): A New Idea for a New Century. http://ciw-online.org/blog/2014/06/wsr/. June 16th, 2014.

CIW website, 2014b. President Bill Clinton, Secretary of State Hillary Clinton honor CIW with Global Citizen Award. http://ciw-online.org/blog/2014/09/bill-clinton/. September 22, 2014.

Colenbrander, A., 2016. International Framework Agreements. In: Utrecht L. Rev.vol. 12. p. 109.

European Federation of Food, Agriculture and Tourism Trade Unions (EFFAT), 2018. Effat.org. (Accessed 30 May 2018).

Ferriss, S., Sandoval, R., 1998. The Fight in the Fields: Cesar Chavez and the Farmworkers Movement. Houghton Mifflin Harcourt.

Gamboa, E., 2000. Mexican Labor & World War II: Braceros in the Pacific Northwest, 1942–1947. 2000, University of Washington Press, Seattle, p. 75.

Mamer, J., Rosedale, D., 1975. Labor management for seasonal farmworkers. In: California Agriculture 29.2. pp. 8–9.

Marosi, R., 2015. Mexican Government Raids Farm Labor Camps in Crackdown on Abuses. In: Los Angeles Times. April 13, 2015.

Mills, F.D., 1964. The National Union of agricultural workers. J. Agric. Econ. 16 (2), 230–258.

Munoz, H.M., 2003. Florida Farmworkers Picket Taco Bell. In: Los Angeles Times. February 25, 2003.

Oxford Business Group, 2018. Cut flower export growth comes as Kenya looks to new markets, August 31. https://oxfordbusinessgroup.com/news/cut-flower-export-growth-comes-kenya-looks-new-markets.

Pawel, M., 2006. UFW: A Broken Contract. In: Los Angeles Times. January 8–11, 2006.

Reardon, T., Timmer, C.P., Barrett, C.B., Berdegue, J., 2003. The rise of supermarkets in Africa, Asia, and Latin America. Am. J. Agric. Econ. 85 (5), 1140–1146.

Riisgaard, L., 2005. International framework agreements: a new model for securing workers rights? Ind. Relat. 44 (4), 707–737.

Riisgaard, L., 2008. Global value chains, labor organization and private social standards: lessons from east African Cut Flower Industries. World Dev. 37 (2), 326–340.

Chapter 8

The End of Farm Labor Abundance

Farmers across California are experiencing the same problem: Seasonal workers who have been coming for decades to help with the harvest, planting and pruning have dropped off in recent years.

SF Chronicle, May 27, 2012

Around the world, as countries develop, their workforces shift out of agriculture. Historically, the United States and other high-income countries relied on immigrant farmworkers from relatively poor countries to fill the void as their own populations left farm work. This chapter presents evidence that the supply of agricultural labor in immigrant-source countries is becoming less and less elastic. What happens when children in farmworker-source countries stop growing up to be farmworkers? Can farmers in high-income countries compete in an era of farm labor scarcity?

For more than a century, agriculture in the United States grew and developed in a context of abundant immigration from less-developed countries. Immigration permitted agricultural wages to remain low as the domestic labor supply transitioned out of hired agricultural work. It limited the effectiveness of union efforts to demand higher wages and improve working conditions. It facilitated the expansion of labor-intensive fruit, vegetable, and horticultural (FVH) production in the absence of large numbers of domestic workers willing to labor in the fields. In short, immigration allowed labor-intensive agriculture to flourish in a country where very few children grow up to be farm workers. The same is true of other high-income countries that have grown to rely on foreign agricultural workers, from Germany to New Zealand.

What if farmers planted crops and the foreign workers did not come?

If the supply of immigrants to farms in high-income countries diminished, agricultural wages would rise. This would put downward pressure on labor demand, as the market adjusted in an effort to realign labor demands with the smaller labor supply.

Agricultural production *could* shift away from labor-intensive crops, but demand for fresh fruits, vegetables, and horticultural commodities rises as

The Farm Labor Problem. https://doi.org/10.1016/B978-0-12-816409-9.00008-2

181

consumers' incomes rise. High-income countries with a smaller farm labor supply, and consequently higher farm wages, could import fresh fruits and vegetables from low-income countries, but this is unlikely to satisfy consumers who have preferences for quality characteristics that include short distances from "farm to fork," farmer accountability, and locally grown produce.

Farmers no doubt would respond by adopting new labor-saving technologies, from "shake-and-catch" machines to robotic strawberry pickers.[1] Engineers at universities and private companies would respond by developing the new technologies that farmers demanded—a process that economists Vernon Ruttan and Yujiro Hayami called "induced innovation" (Ruttan and Hayami, 1984). Farms would employ fewer workers, but these workers would have to possess the skills to use and maintain the new technologies.

In fact, changes like these currently are underway in the United States. Rural Mexicans are transitioning out of agricultural work, and the supply of immigrants to hired farm work in the United States is shifting inward. This has far-reaching implications for the food farmers grow and how they grow it. The same process is underway in other high-income countries around the world, as well as in many not-so-high-income ones (Taylor, 2010).

In Chapter 4, we saw that regional changes in the demand for labor due, for example, to a late harvest or high-yield year, can create localized labor shortages. Local labor shortages are not uncommon, but labor shortages on US farms have become more widespread and appear to be a national, rather than local, phenomenon. This raises the question: Has there been an inward shift in the immigrant farm labor supply affecting all regions of the country?

Rural Mexico supplies the vast majority of hired labor to US farms. Is rural Mexico transitioning out of farm work as the Arthur Lewis model in Chapter 3 would predict? How can we test this hypothesis?

Taylor and Lopez-Feldman (2007) used household survey data to measure the impact of migration on agricultural productivity in rural Mexico. They found that migration to the United States increased household income in Mexican villages, both through remittances and through increased agricultural productivity. It appears that when one member of an agricultural household migrates, the marginal productivity of labor by the remaining household members rises. This suggests that rural Mexico has passed the Lewis turning point, at which industries compete with farms for a limited supply of labor.

That study does not tell us how quickly rural Mexicans are leaving agricultural work or the implications for the US agricultural industry. For that, we need to measure the probability that individuals from rural Mexico do farm work, whether in the United States or Mexico, over an extended time period during which the US demand for farm labor remained steady or increased.

---

1. Shake-and-catch machines surround the trunks of trees and shake the harvestable fruit into catching frames. These are used primarily for nut harvest since nuts do not bruise, although farmers have also found applications of shake-and-catch methods in the harvest of other less delicate fruits as well.

If the share of individuals from rural Mexico working in agriculture decreases even as the demand for hired agricultural workers increases, this would suggest that there was an inward shift in the farm labor supply. Taylor et al. (2012) conducted a "quasiexperimental" analysis with household survey data nationally representative of rural Mexico before and after the Great Recession of 2008. A quasiexperiment, like an experiment, tests for changes in an outcome (here, the probability of doing farm work) before and after some exogenous or unexpected change, or "treatment" (here, the Great Recession of 2008).

During the Great Recession, the demand for labor in the US nonfarm sector declined while the demand for labor in the agricultural sector remained steady. For example, when the housing crisis hit, the demand for construction workers tanked. Before the recession, some farmers complained that building contractors were luring their workers away to build houses in California's booming Central Valley cities, from Bakersfield to Sacramento. In theory, the recession should have reduced Mexican workers' probability of working in construction, while increasing their supply of labor to farms. If the probability of doing nonfarm work fell, but so did the probability of doing farm work, that would suggest that something else was happening—like a reluctance by immigrants to do farm work. We will look at the results of this quasiexperiment in more detail later in this chapter, along with a more in-depth analysis by Charlton and Taylor (2016) that uses econometrics to measure the trend in the probability of working in agriculture from rural Mexico between 1980 and 2010.

This chapter reviews evidence that relatively poor countries that supply immigrant workers to farms are transitioning out of farm work, just as the United States and other high-income countries did in the mid-20th century. The agricultural industry in high-income countries has few options going forward. It can shift away from labor-intensive crops, but as long as the demand for locally grown fruits and vegetables remains steady this will not be a viable option. Farms can seek immigrants from other countries, but the movement of people out of farm work is becoming a global phenomenon, just as the Lewis model predicted. This leaves us with the question of where the new workers would come from (and also whether governments would admit large numbers of immigrants from far-flung places like China and India) to do farm work.

Alternatively, farmers can invest in technologies that increase the marginal product of labor, leading to higher wages and more skilled jobs. Investing in labor-saving agricultural technologies appears to be the most viable option for the agricultural sector in high-income countries. It can offer potential long-term benefits for farm workers and rural communities (Chapter 9).

As the immigrant farm labor supply becomes more inelastic and wages rise, the circular relationship between farm labor demand, immigration, and rural poverty documented by researchers in the 1990s and early 2000s is expected to come to an end.[2] Farm workers, the families of farm workers, and the

---

2. For example, see Martin et al. (2006).

communities where farm workers live will benefit from rising agricultural wages. Rural communities will experience economic growth, as more skilled and better-paid agricultural employees increase their spending and standards of living rise. In theory, the transition of labor away from agriculture can be positive for rural communities in high-income countries, like the United States, as well as immigration-source countries, like Mexico. The challenge for the agricultural industry will be to anticipate and adjust to a changing farm labor market.

WHEN THE FARM LABOR SUPPLY BECOMES LESS ELASTIC

In 2000, rural Mexico supplied 79% of the hired agricultural labor employed in the United States (United States Department of Agriculture, 2016). Consequently, changes in the farm labor supply from rural Mexico can have severe impacts on the agricultural labor market in the United States. We can illustrate how changes in the farm labor supply from rural Mexico are likely to impact the agricultural labor market in the United States using a diagram showing interconnected labor markets.

Fig. 8.1 illustrates interconnected farm labor markets between Mexico and the United States. Panel a illustrates the supply and demand for farm labor in the United States and panel b illustrates the supply and demand for farm labor in rural Mexico.

Initially, the agricultural labor supply in rural Mexico is very elastic, illustrated by LS_{MX}. Workers from rural Mexico are willing to migrate to US farms as long as the wage in the United States is at least as high as the equilibrium wage in Mexico plus the cost of migrating. Let δ be the cost of migration.

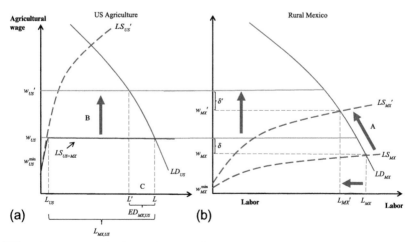

FIG. 8.1 Interconnected US and Mexican agricultural labor markets. As the farm labor supply from rural Mexico becomes more inelastic (arrow A in panel b), the reservation wage for Mexican immigrants to US farms rises (arrow B in panel a). *(From Taylor, J.E., Charlton, D., Yúnez-Naude, A., 2012. The end of farm labor abundance. Appl. Econ. Perspect. Policy 34(4), 587–598.)*

US farms can hire up to L workers at wage $w_{US} = w_{MX} + \delta$ *(panel a)*. This is the reservation wage, or minimum wage required to induce a worker from rural Mexico to migrate to agricultural work in the United States. LS_{US} domestic workers will be employed on the US farms at this wage, and $L - LS_{US}$ immigrants will come from rural Mexico to work on the US farms.

If rural Mexico begins to shift out of agricultural work, then the farm labor supply in Mexico will rotate inward and become more inelastic, as illustrated by the shift from LS_{MX} to LS_{MX}' in panel b. With a more inelastic farm labor supply, the agricultural wage in rural Mexico must rise from w_{MX} to w_{MX}', and the reservation wage to pull workers from rural Mexico to farms in the United States must rise from w_{US} to w_{US}'. The number of workers employed on US farms at the new equilibrium drops from L to L'. If farms are slow to adjust to the new equilibrium, there will be an excess demand for farm workers at below-market wages.

If our hypothesis that rural Mexico is shifting out of agricultural work is correct, then we would expect to see a simultaneous rise in farm worker wages in both the United States and Mexico, just as the graphs in Fig. 8.1 predict. In fact, farm wages in the United States are rising. Fig. 8.2 illustrates the change in farm worker wages between 2011 and 2013 by region. In nearly every region of the United States, farm worker wages rose much faster than the consumer price index (CPI). Because of this, real (inflation-adjusted) wages increased.

Agricultural wages are rising in Mexico as well. The Mexican government does not gather detailed wage data; however, anecdotal and newspaper reports

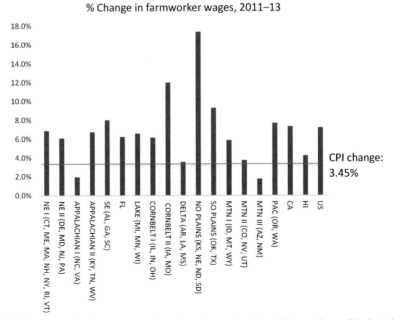

FIG. 8.2 Agricultural wages are rising throughout the United States. *(From Charlton, D., Taylor, J.E., 2016. A declining farm workforce: Analysis of panel data from rural Mexico. Am. J. Agric. Econ. 98(4), 1158–1180.)*

indicate that agricultural wages in Mexico are rising. Miguel Curiel, Deputy General Manager of Driscoll's in Mexico, reports a 92% increase in wages on berry farms from 2016 to 2018, from P$130 to P$250 per day for full-day workers.[3] Worker strikes in Baja Mexico in 2015 ended in wage increases of around 15% along with other improvements in working conditions and benefits (Marosi, 2015). One of the challenges to farm labor organizing is that strikes are only successful when there is not an elastic supply of labor in reserve to replace the workers on strike (see Chapter 7). The success of strikes in Baja Mexico suggests that the farm labor supply is becoming less elastic.

Rising farm wages are not sufficient evidence to prove that rural Mexico is transitioning out of agricultural work. Other factors could cause the same phenomenon. Imagine, for example, the impact of an outward shift in the demand for farm workers when the farm labor supply is not perfectly elastic. To determine whether the farm labor supply from rural Mexico has shifted inward, we need to consider a more experimental design. We need to see what happens to farm employment when the demand for farm labor is constant. Taylor et al. (2012) took advantage of a quasinatural experiment of this nature during a recession that negatively impacted only the nonfarm industry. Let us look at their experimental design and findings in more detail.

WAS THERE A SHIFT IN THE FARM LABOR SUPPLY? A QUASINATURAL EXPERIMENT

There have been numerous reports of farm labor shortages in the newspapers, and data indicate that farm wages are on the rise. Both of these observations could be caused by an inward rotation of the farm labor supply from rural Mexico, but we need to be certain that the farm labor shortages reported in the United States are not caused by temporary disequilibria in local markets and that rising wages were not caused by a shift in the demand for farm labor.

We might be able to determine whether reported farm labor shortages reflect a national phenomenon if we could observe the total number of hired workers employed in agriculture at the national level, or if we could observe the share of rural Mexicans employed in agriculture in either Mexico or the United States at different points in time. In fact, we can observe the share of individuals from rural Mexico employed in agriculture using data from the National Rural Mexican Household Survey (Spanish acronym ENHRUM). This survey is nationally representative of rural Mexico, and it interviewed the same households in 3 years: 2002, 2007, and 2010. It asked where every member of the household, including all children of the household head and his or her spouse, worked the previous year—whether in Mexico or the United States, whether in agricultural or nonagricultural jobs, and whether self-employed or working for a wage.

---

3. Email correspondence, May 24, 2018. The peso-to-dollar exchange rate varied between 18 and 21 during this period.

Taylor, Charlton, and Yúnez-Naude (2012) used these data to determine whether the share of individuals working in the agricultural sector declined between 2002 and 2007, and again between 2007 and 2010. The years from 2007 to 2010 are unique, because they sandwich a large, unexpected recession that began in 2008. The recession was concentrated in the US nonfarm industries (construction was particularly hard hit), but other countries that trade with the United States, including Mexico, were also impacted. By 2010, the economy was on the path to recovery, but it had still not rebounded to its pre-2008 state.

The Great Recession of 2008 provides a quasinatural experiment for studying changes in the probability that rural Mexicans work in the agricultural sector, because individuals chose where to work in 2007 without knowing or expecting that the recession would come in 2008. During the recession, the demand for nonfarm workers shifted inward, but the demand for agricultural workers remained steady. People continued to eat even when the recession hit. Moreover, a large share of farmland in California, where the demand for farm labor is highest, is in tree orchards and vineyards that must be maintained no matter what the rest of the economy is doing.

We might expect migrants working in the US nonfarm sector prior to 2008 to seek jobs in the agricultural sector, where the demand for labor persisted after the recession hit. However, this is not what Taylor et al. (2012) observed. They found a decline in migration from rural Mexico to *both* the farm and nonfarm sectors in the United States between 2007 and 2010. Not only that, the percentage decrease in migration to agricultural jobs was *larger* than the percentage decrease in migration to nonagricultural jobs (Fig. 8.3). More individuals switched from US farm to nonfarm work between 2007 and 2010 than the reverse. That is the opposite of what we would expect to observe if there was only a shift in the demand for nonfarm workers between 2007 and 2010. The findings from this study, which won an award from the Agricultural and Applied Economics Association (AAEA), suggest that there was a simultaneous downward shift in the supply of rural Mexican workers.

This study offered the first empirical evidence that rural Mexicans were transitioning out of agricultural work, even as the demand for nonfarm workers in the United States shifted inward. This implies that immigration policies alone cannot resolve problems of farm labor shortages in the United States.

HOW QUICKLY IS THE WORKFORCE FROM RURAL MEXICO SHIFTING OUT OF AGRICULTURAL WORK? AN ECONOMETRIC ANALYSIS

This quasinatural experiment involving the Great Recession of 2008 suggests that rural Mexico is transitioning out of agricultural work, but a big question remains: How quickly is the transition happening? Should US farmers pressure Congress to enact a guest worker program in response to farm labor shortages? If it did enact a guest worker program with Mexico, would the farmworkers come? Would farmers be better off offering higher wages and better benefit

Percentage change in number of Rural Mexicans working in each sector-location

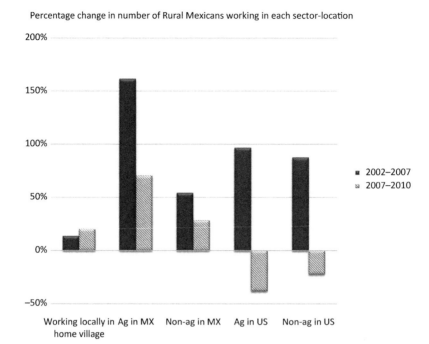

FIG. 8.3 After the Great Recession hit in 2009, the share of rural Mexicans working in US non-farm jobs fell, but the share in farm jobs fell more. *(From Taylor, J.E., Charlton, D., Yúnez-Naude, A., 2012. The end of farm labor abundance. Appl. Econ. Perspect. Policy 34(4), 587–598.)*

packages to retain existing workers? How do other factors, like increased border enforcement, affect the supply of Mexican workers to US farms? Would a change in immigration policy or border enforcement help?

As applied economists, we can seek answers to these questions by analyzing additional work history data. The ENHRUM collected work histories for all members of every household in the random sample every year from 1980 to 2010, including whether they were working in agricultural jobs in Mexico or the United States. This amounts to 31 years of data on 9,918 individuals who were of working age at some point in the 1980–2010 period, approximately 155,000 person-years of data.[4] Econometricians refer to data on many individuals over time as a *matched panel*.

We can use matched panel data to track individual workers into and out of farm work over this extended period of time. The time period covered by our data, more than two decades, spans almost a generation of lives, many economic ups and downs in Mexico and the United States, increases in border enforcement, and changes in many other variables that might influence the migration

---

4. Knowing where one person worked in each of 5 years provides 5 person-years of data, knowing where 5 people worked in each of 20 years provides 100 person-years of data, and so on.

of rural Mexicans to US farms. We can use these data to identify a trend in the probability that an individual from rural Mexico works in agriculture in the United States or Mexico, and to determine what variables explain this trend.

We can begin by looking at some summary statistics from the matched panel. Table 8.1 shows the share of individuals in distinct age groups that worked in agriculture in the United States or Mexico in 1980, 1990, 2000, and 2010.

TABLE 8.1 Share of Individuals Working in Agriculture in Mexico and the United States by Age

Age	Location-Sector	Mean	SD	Min	Max	Obs
1980						
15–29	US Agriculture	0.012	0.107	0	1	1213
	MX Agriculture	0.384	0.487	0	1	1213
30–39	US Agriculture	0.010	0.010	0	1	500
	MX Agriculture	0.500	0.501	0	1	500
40–49	US Agriculture	0.006	0.080	0	1	309
	MX Agriculture	0.576	0.495	0	1	309
50–65	US Agriculture	0.005	0.069	0	1	208
	MX Agriculture	0.611	0.489	0	1	208
1990						
15–29	US Agriculture	0.016	0.127	0	1	2573
	MX Agriculture	0.316	0.465	0	1	2573
30–39	US Agriculture	0.027	0.161	0	1	972
	MX Agriculture	0.355	0.479	0	1	972
40–49	US Agriculture	0.020	0.140	0	1	601
	MX Agriculture	0.434	0.496	0	1	601
50–65	US Agriculture	0.002	0.047	0	1	449
	MX Agriculture	0.530	0.500	0	1	449
2000						
15–29	US Agriculture	0.024	0.154	0	1	3069
	MX Agriculture	0.246	0.431	0	1	3069
30–39	US Agriculture	0.025	0.156	0	1	1567
	MX Agriculture	0.266	0.442	0	1	1567

Continued

TABLE 8.1 Share of Individuals Working in Agriculture in Mexico and the United States by Age—cont'd

Age	Location-Sector	Mean	SD	Min	Max	Obs
40–49	US Agriculture	0.032	0.176	0	1	972
	MX Agriculture	0.332	0.471	0	1	972
50–65	US Agriculture	0.016	0.124	0	1	833
	MX Agriculture	0.438	0.496	0	1	833
2010						
15–29	US Agriculture	0.010	0.101	0	1	2058
	MX Agriculture	0.185	0.388	0	1	2058
30–39	US Agriculture	0.016	0.124	0	1	1341
	MX Agriculture	0.181	0.385	0	1	1341
40–49	US Agriculture	0.017	0.128	0	1	1014
	MX Agriculture	0.255	0.436	0	1	1014
50–65	US Agriculture	0.018	0.131	0	1	802
	MX Agriculture	0.298	0.458	0	1	802

From Charlton, D., Taylor, J.E., 2016. A declining farm workforce: Analysis of panel data from rural Mexico. Am. J. Agric. Econ. 98(4), 1158–118.

You should notice two patterns in this table. First, in any given year, older people are more likely to do farm work. The top panel of the table shows where rural Mexicans worked in 1980. About 39% of workers 15–29 years old did farm work (1.2% in the United States and 38.4% in Mexico). In the 50–65-year age group, nearly 62% did farm work in 1980 (61.1% in Mexico and only 0.5% in the United States). We see a similar pattern by comparing young and old workers in each of the other 3 years shown in the table.

The second pattern to notice is that, for the same age group, the likelihood of working in agriculture decreases overtime. We saw that around 39% of the 15–25 age group did farm work in 1980. This share drops to below 20% in 2010, when 1% of this age group did farm work in the United States and 18.5% did farm work in Mexico. Of workers in the oldest age group, around 62% did farm work in 1980, but just over 30% did farm work in 2010 (1.8% in the United States and 29.8% in Mexico).

In short, no matter how you look at it, younger workers are less likely to do farm work than older workers, and everyone of the same age is less likely to do farm work in 2010 than in 1980. The decline in rural Mexicans' probability of doing farm work over time is striking.

To quantify the downward trend in the probability of working in agriculture, Charlton and Taylor (2016) estimated a dynamic linear regression model (see Box 1). In their model, the dependent (left-hand) variable, Y_{it}, is an indicator variable equal to 1 if person i did farm work in year t and zero otherwise. You can think of this dependent variable as representing the probability of doing farm work, like in the mover-stayer model of Chapter 3. The right-hand

---

BOX 1 What's a Linear Regression?

Economists often use real-world data to estimate relationships or correlations between one variable (say, X) and another (Y). The line in the figure below is a linear regression of Y on X. The variable X might be a year number, ranging from 0 (the base year of the data) to T (the final year). Y might be a measure of farm work, like the number of days a person worked in agriculture during that year, or simply a 1–0 indicator of whether or not a person did farm work at some time during the year. The slope of the line would then be the estimated change in farm labor per year, or the time trend. For example, if the dependent variable is days worked and the slope is −4, as in this made-up illustration, people work 4 fewer days in agriculture from 1 year to the next on average. In Charlton and Taylor's study, the dependent variable is an indicator or "dummy" variable equal to 1 if a person worked primarily in agriculture in a given year and 0 otherwise. The regression line they estimated has a slope of approximately −0.01. This means that, on average, the probability of doing farm work decreased by 0.01 (or 1%) per year.

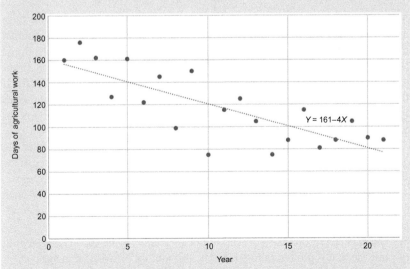

Hypothetical example of a negative trend in days of agricultural work per year

Econometrics is about finding the best way to model relationships, which usually involve not just two but many variables (Smith and Taylor, 2017).

(explanatory) variables include an annual trend (t), which is simply the year number (1 for 1980, 2 for 1981, and so on), and an indicator of whether the individual worked in agriculture in each of the previous 2 years (Y_{it-1} and Y_{it-2}). It also includes an error, $\varepsilon_{i,t}$. In short, the model looks like this

$$Y_{i,t} = \beta_0 + \beta_1 t + \gamma_1 Y_{i,t-1} + \gamma_2 Y_{i,t-2} + \varepsilon_{i,t}$$

In this equation, β_1 measures the change in probability of doing farm work from 1 year t to the next, which is the time trend we wish to estimate. If β_1 is positive, the probability of working in agriculture is increasing, and if it is negative (as we suspect), it is decreasing. Including variables indicating whether individuals worked in agriculture in each of the previous 2 years controls for autocorrelation, or persistence in the probability that an individual continues working in the same sector from year to year. It is important to control for this in a model with repeated observations on each individual over time.

Charlton and Taylor (2016) found that rural Mexicans' probability of working in agriculture is declining at a statistically significant rate of 0.97 percentage points each year. (The estimate of β_1 is -0.0097.) That may sound like a small change, but over many years it represents a substantial decline in the farm labor supply. Scaling by the working-age population of rural Mexico in 2010, their findings imply that the farm labor supply from rural Mexico is decreasing by more than 150,000 people annually.

A REGION-BY-REGION LOOK AT THE DECLINING FARM LABOR SUPPLY

Mexico is a large and diverse country. One might suspect that trends in the farm labor supply vary from one region in Mexico to another, and if so, internal migration could fill the void in labor-shortage regions. Data from the National Agricultural Worker Survey (NAWS) show that the share of the US farmworkers from southern Mexico has been increasing (Gabbard et al., 2015). If the farm labor supply is drying up in some regions of Mexico, could increases in other regions compensate for this and ensure a future supply of workers for US and Mexican farms?

We can test whether trends in farm labor supply vary across regions in Mexico. The ENHRUM was designed to be representative of rural households in each of the five census regions of Mexico, as well as of all rural households in the country.

Charlton and Taylor (2016) estimated separate linear regressions for each census region of Mexico. They found a significant negative trend in the probability of working in agriculture in *every* census region.

The expected probability of working in agriculture from each census region of rural Mexico is plotted in Fig. 8.4. The probability of working in agriculture is higher for individuals from more southern regions of Mexico, indicated by higher intercepts for Central, West-Central, and South-Southeast Mexico. Nevertheless, there is a steep downward trend in the probability of working in

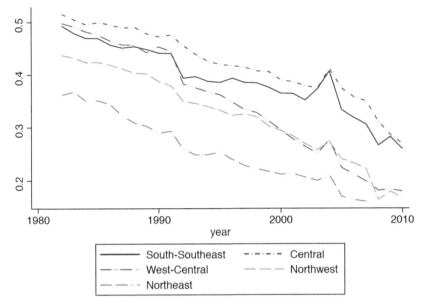

FIG. 8.4   The probability of working in agriculture is declining in every Mexican census region.

agriculture in every region. The rural workforce is transitioning out of agricultural work in all regions of Mexico.

WHY DOES THE FARM LABOR SUPPLY BECOME MORE INELASTIC?

The analysis described in this chapter provides strong evidence that rural Mexico is transitioning out of agricultural work. Why is this happening? Should we expect the downward trend to continue in the future? Farmers across the United States ask whether the supply of Mexican workers will return to what it once was. Whether or not it will recover depends on what factors are causing the downward trend and whether they are likely to change. More generally, why does the farm labor supply decrease as countries become more economically developed? The farm labor transition underway in Mexico can provide insights into this important question.

To make a model to uncover the determinants of the declining farm labor supply, we need to return to some of the economic theory we discussed in previous chapters. In Chapter 3, we looked at the Lewis model, which posits that competing demand for workers in the agricultural and nonagricultural sectors drives the agricultural transition. We might investigate whether an expanding nonfarm sector competes with agriculture in Mexico, pulling workers off the farm. We could create a model that controls for changes in nonfarm employment in Mexico each year to measure the impact of nonfarm employment on the probability that individuals work in agriculture.

How do individuals decide which sector to work in when there are competing demands for their labor? The mover-stayer model posits that the characteristics of an individual and his or her household impact the decision either to remain in agriculture or move to the nonfarm sector. The number of able-bodied people in the household might determine whether an individual migrates to the nonfarm sector, given decreasing marginal productivity of labor on farms. The birthrate in rural Mexico fell sharply in the late 20th century.

Human capital is another important element impacting individuals' sector of work in the mover-stayer model. Education in rural Mexico is increasing. The average education of an individual in his or her 50s in 2010 was 5 years, or just shy of the completion of primary school in Mexico, according to the ENHRUM data. In contrast, the mean education of an individual in his or her 20s was 9 years, or the completion of lower-secondary school. How does increased education affect the probability that individuals work in agriculture?

The law of supply and demand implies that farm wages impact whether individuals remain in agriculture. Consequently, we should control for farm wages and the US-Mexico exchange rate, which converts dollar earnings in the United States into pesos in Mexico. Many individuals migrate to work in the United States. The intensity of US border enforcement directly affects migration costs and risks, and thus the reservation wage of migrating. We can control for changes in both US farm wages and the number of US border patrol agents over time.

Ideally, we would like to control for changes in Mexican farm wages, but the availability of Mexican wage data limits us. Theoretically, US and Mexican farm wages should be highly correlated, since a porous border separates the two farm labor markets. High costs of migrating would explain why farm wages are higher in the United States than in Mexico, but given how connected the two labor markets are, one would expect US and Mexican farm wages to move in the same direction. Once we control for US farm wages, therefore, Mexican farm wages are likely to have little impact on the farm labor supply.

Charlton and Taylor added all of these variables to their econometric model. The new model regresses the same 1–0 indicator variable measuring whether or not person i worked in agriculture in year t on an annual trend, the individual's age, gender, years of education, ratio of children to adults in the household, and lagged and differenced US farm wages, number of border patrol agents, the Mexican peso-US dollar exchange rate, employment in the Mexican service sector, employment in the Mexican industrial sector, and homicide rate in the individual's home municipality.[5]

The model also includes an indicator variable for each household, which economists call "fixed effect." By controlling for household fixed effects, the model identifies the respective impact of each factor on changes in the probability that individuals within *the same household* work in agriculture. It therefore eliminates any differences in the probability of working in agriculture that

---

5. The coefficient on the homicide rate was not significantly different from zero, so its omitted from Figure 8.5.

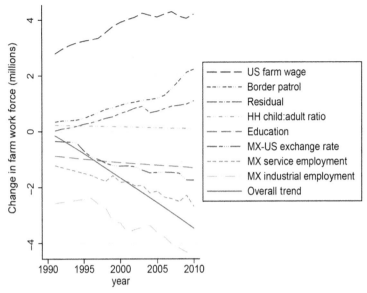

FIG. 8.5 Growth in nonfarm jobs, rising education, and falling birthrates in Mexico are driving the decrease in farm labor supply. Each *dashed or dotted line* illustrates the impact of individual factors on the agricultural workforce from rural Mexico. The sum of all impacts is the overall trend, marked by the *downward-sloping solid line*. (*From Charlton, D., Taylor, J.E., 2016. A declining farm workforce: Analysis of panel data from rural Mexico. Am. J. Agric. Econ. 98(4), 1158–118.*)

might be explained by time-invariant household characteristics, including ones that are not possible to measure.

Fig. 8.5 plots the relative impact of each of these factors on the farm labor supply from rural Mexico from 1980 to 2010. To translate impacts on the *probability* of doing farm work, estimated in the econometric model, to changes in the farm labor *supply*, we multiply the estimated probability of working in agriculture by Mexico's working-age population each year. The curves below the horizontal line at zero correspond to variables that accelerate the negative trend. The curves lying above the zero line correspond to variables that counteract the negative trend.

These curves tell us the change in number of rural Mexicans working in farm jobs that is attributable to the corresponding variable. For example, in 1990, Mexican industrial employment accounted for around 2.5 million fewer rural Mexicans employed in farm work each year (see *dashed* and *triple dotted line* at the bottom of the figure). By 2010, it accounted for more than 4 million fewer. The overall trend is given by the downward-sloping solid line near the bottom of the figure, and it shows the change in the farm labor supply since 1990.

Nonfarm employment, peso devaluations, and rising education in rural Mexico are all significant factors moving rural Mexicans out of farm work. Nonfarm jobs compete with agriculture to attract workers from rural areas. Schooling prepares young rural Mexicans to take on nonfarm jobs.

Peso devaluations have potentially complex effects. They raise the value of dollars earned in the United States and sent back to Mexico. This could make working in the United States more attractive compared with working in Mexican agriculture, although this argument could apply to the US nonfarm as well as farm jobs. (Most Mexican migrants in the United States work in nonfarm jobs.) Peso devaluations also make Mexico's exports, most of which are non-agricultural goods, cheaper to foreign consumers. This could stimulate the nonfarm sector in Mexico, moving people off the farm.

All of these factors appear to increase the opportunity cost of working in farm jobs for people from rural Mexico. As the opportunity cost of doing farm work rises, the mover-stayer model tells us that people become more likely to shift out of farm work.

A number of variables partially counteract the negative trend. Rising US farm wages increase the number of people doing farm work. However, the impact of rising farm wages is not large enough to reverse the downward trend in the farm labor supply.

Intensified border enforcement also slows the movement of rural Mexicans out of agriculture, no doubt by helping to "trap" them on the farm in Mexico. It appears that many migrants who would have worked in the nonfarm sector in the United States do farm work in Mexico if they are unable to cross the border.

The ratio of children to adults in rural Mexican households declined between 1990 and 2010. It negatively impacts the probability of working in agriculture if we allow variation between households, but (as shown in this figure) it has little impact on the farm labor supply after controlling for household fixed effects. The residual trend includes the impacts of age, gender, and many other unobserved factors that contribute to the decision to work in the agricultural sector.

Isolating the factors that drive the transition of labor out of agriculture is critical if we want to know whether the negative trend is likely to persist. Education is a particularly important factor transforming a nation's workforce because it builds human capital, and the economic return to human capital is typically higher in the nonfarm sector. Increases in education tend to become self-perpetuating, persisting over multiple generations. Once a school is built near a village, children from one generation to the next can utilize the school. Growth in the number of primary and secondary teachers in Mexico reveals that education is becoming more available. Studies show that once people with more education become parents, they invest more in the education of their own children (Handa, 2002).

Analysis of the factors contributing to the declining probability of working in agriculture from rural Mexico indicates that rural Mexico is undergoing a structural shift that is not likely to reverse itself. As opportunities in the nonfarm sector of Mexico expand, rural education rises, and birthrates decline, the opportunity cost of working in agriculture will continue to increase. Farm wages are rising, and the number of individuals working in the agricultural sector is declining, just like the Lewis model predicts will happen as an economy develops.

BEYOND MEXICO AND THE UNITED STATES: A GLOBAL TREND

Nowhere in the world do researchers have access to the same detailed data to study changes in the farm labor supply as in Mexico and the United States. Nevertheless, the data that are available provide convincing evidence that the declining farm labor supply is a worldwide phenomenon, consistent with the predictions of the Lewis model. Workforces are becoming less agricultural throughout the North American region. Fig. 8.6A shows changes in per capita income and the farm workforce share for Mexico, Guatemala, El Salvador, Honduras, and the United States. The origin of each ray in this figure shows per capita income and the farm workforce share in 1998; the tip of the ray shows the same in 2006. Both the slope and positions of the rays reveal that the share of the workforce in agriculture falls precipitously as incomes rise. The only exception to this trend is the ray corresponding to Honduras, which is very short and indicates a slight increase in the farm workforce share. Nevertheless, its position at the top of the workforce transition curve reaffirms the negative correlation between per capita income and workforce share.

The share of the total workforce employed in agriculture is high in Mexico and Central America relative to the United States, but it is falling fast. Across Mexico and Central America, educational attainment is increasing and incomes are rising, although these advances and demographic trends are evolving at different speeds in each country. Mexico and El Salvador are seeing their populations age and demographic growth slow (Terrazas et al., 2011). In contrast, birthrates remain relatively high in Guatemala and Honduras. As young people in these countries attain higher levels of education, they will seek employment opportunities beyond domestic agricultural work, migrating away from rural areas and, within rural areas, leaving farm jobs for jobs in the service or manufacturing.

In 2011–12, approximately 28,000 Guatemalans annually, the vast majority of them farm workers, were issued multiple-entry, year-long border cards to work in Mexico (Instituto Nacional de Migración (INM), 2011; INM, 2012). It is difficult to estimate how many Guatemalans are employed in agriculture in Mexico overall, as the border between the two countries is porous, undocumented entry and employment are widespread, and work is seasonal and often short term, leading to multiple entries. A survey of Central American migrants crossing the Mexico-Guatemala border estimated 153,000 border crossings on average per quarter by Guatemalans going to work in Mexico between the second quarter of 2011 and the first quarter of 2012. In all, 48%–58% of migrants returning to Guatemala after working in Mexico during this period reported working in agriculture, depending on the time of year; upwards of 79% reported being undocumented (El Colegio de la Frontera Norte et al., 2012).

Taken together, these data suggest an incipient tension between agricultural labor supply and demand in the North American region. On one hand, labor-intensive FVH production is expanding in the four countries of Mexico, Guatemala, El Salvador, and Honduras. On the other hand, as incomes increase, the

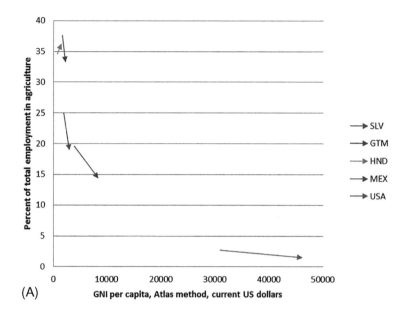

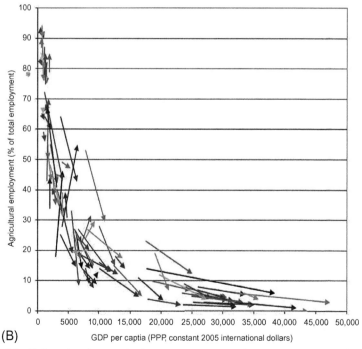

FIG. 8.6 Trajectories of per capita income and the share of labor in agriculture, 1998–2006. (A) Mexico, Guatemala, Honduras, El Salvador, and the United States and (B) international comparison. *(From Taylor, J.E., Charlton, D., Yúnez-Naude, A., 2012. The end of farm labor abundance. Appl. Econ. Perspect. Policy 34(4), 587–598.)*

workforce shifts out of agriculture. The United States marks the extreme of the downward trend in domestic farm labor supply: in 2010 only 1.6% of the US workforce was employed in agriculture, and a significant majority of hired farm workers were Mexican.

When it comes to being vulnerable to a declining foreign farm workforce, US farmers are not alone. High-income countries around the world depend on foreign workers to harvest their crops. Canada imports farmworkers from Mexico and the Caribbean. Germany, the United Kingdom, and Ireland all depend heavily on farm workers from Poland. France employs large numbers of agricultural workers from its former colonies in Africa. Australia and New Zealand import farm workers from the Pacific Islands, including Tonga, Vanuatu, and Kiribati. South Africa, from Zimbabwe and Lesotho.

At the same time, workers are moving off the farm worldwide (Fig. 8.6B). As per capita incomes rise, people move out of farm work and into nonfarm jobs. In the United States, this process has pretty much run its course. Mexico's per capita income (adjusted for the cost of living) was $17,740 in 2017. Mexico is entering the later stages of its farm labor transition.

A declining farm labor supply induces farms to adopt labor-saving technologies that raise the productivity of agricultural workers. In Mexico, total agricultural production is rising, while total farm employment is falling. Because of this, the average productivity per farm worker has risen dramatically. Farm worker productivity more than tripled between 1995 and 2009 (see Fig. 8.7).

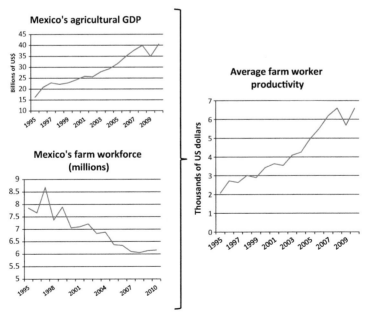

FIG. 8.7 Average farm worker productivity, the agricultural GDP divided by the size of the farm workforce, rose in Mexico from 1995 to 2010. *(Data from FAOSTAT, Food and Agriculture Organization of the United Nations (FAO), http://faostat3.fao.org/home/index.html#DOWNLOAD.)*

FIG. 8.8   Farmworker wages increased rapidly in California. *(Data from California Employment Development Department.)*

Increased productivity means that Mexican farm workers have a higher agricultural reservation wage, making them more likely to stay at home rather than emigrating to work in agriculture abroad. Adjusted for inflation, the average daily wage in the Mexican agricultural sector rose nearly 14% from 2000 to 2007, and if the wages of berry workers reported earlier in this chapter are any indication, the trend has accelerated more recently.[6] Meanwhile, across the border in California, farm wages increased at what almost resembles an exponential rate between 2002 and 2014 (Fig. 8.8), and they continued this upward trend in recent years.

FARMERS' OPTIONS IN AN ERA OF LABOR SCARCITY

Confronted by rising wages and fewer farm workers, farmers' strategies to respond to the end of farm labor abundance can take either of two broad forms.

The Exploration and Development Response

In theory, farmers could respond to a diminishing labor supply by seeking new labor sources. In the past, agriculture in high-income countries sought out new supplies of farm labor to relieve labor shortages and limit increases in farm worker wages. This response is akin to what is called the "development

6. Calculated using data from the Mexican Social Security Institute, available in UN-ECLAC (2011).

response" in the case of natural resources. For oil, it involves exploration. For farm labor, since the days of the Bracero Program, it has involved immigration policy to ease farmers' access to foreign workers.

Securing new supplies of farm labor requires labor recruiters and migrant networks to extend their reach into new farm worker-source areas. For example, during the Bracero Program, labor recruiters tapped surplus workers in Mexican villages close to railroad lines for easy transport to the US-Mexican border. By the end of the Bracero Program in 1964, labor recruiters no longer were needed; informal networks connecting villagers in Mexico with friends and family members in the United States took the recruiters' place as the driving force behind the farm labor supply.

Similar networks appear to drive migration from Poland to farms in the United Kingdom and Germany, from Albania to Greece, Guatemala to Mexico, Nicaragua to Costa Rica, and many other migrant-source to migrant-destination countries. Where international farm labor flows are largely unauthorized, human smugglers constitute a vital link in the international migration chain, strategically and surreptitiously moving thousands of workers across boarders—for a price. Labor recruiters play an important role in legal international migrant worker movements from Asia and Africa to the oil-rich Middle East (Massey and Taylor, 2004).

The immigration solution is a viable option as long as the supply of labor to farm work in the labor-source area is high. That is why, even though there is free movement of labor across US states, interstate migration does not supply significant numbers of domestic workers to US farms anymore, and why economic integration with the European Union did not solve the UK's farm labor problems until Poland and other relatively low-wage countries to the east also joined the European Union. There are limits to the effectiveness of immigration policy in an era when the supply of labor to farm work is declining in labor-source countries.

The US H-2A agricultural guestworker program has expanded rapidly in an era of diminished farm labor supply. The US government certified 165,741 H-2A jobs in 2016. This represented 160% growth since 2006, when only 64,100 jobs were certified (Martin, 2017). Growth in demand for H-2A workers reflects farmers' difficulties finding seasonal farmworkers through local job advertising and worker networks, consequences of increased competition with Mexican farms and nonfarm jobs. Although H-2A visas might make migrating to US agriculture more attractive, the number of workers willing to come to US farms even with a guestworker visa is expected to decline in the long run as Mexicans continue their transition out of farm work.

Central American migration to US farms rose from 2% of all crop workers in federal fiscal years 1999–2000 to 4% in 2013–14 (US Department of Labor, Employment and Training Administration, 2016). With a total farm workforce only one-fourth the size of Mexico's and a declining farm employment share of its own, however, Guatemala is unlikely to have more than a marginal effect on the US farm labor supply. Moving further south, Honduras and El Salvador have small work forces compared with Guatemala, and the shares of agriculture in total employment are falling faster there than in Guatemala.

The rural populations of Asia and Africa are relatively large, but contracting hundreds of thousands of farm workers from more distant countries is costly, and gaining political support for this does not appear feasible. Furthermore, the factors pushing and pulling Mexicans out of farm work, such as rising education and a growing nonfarm economy, are at play in other developing countries.

The UK's decision to leave the European Union has major implications for UK farmers' access to farmworkers, because many Eastern Europeans, particularly from Poland, currently supply their labor to UK farms. The Food Supply Chain manifesto, sent to Prime Minister Theresa May by National Farmers' Union president Minette Batters in May 2018, argues that, due to the "significant number" of EU nationals working in UK fields, it is vital that the government "ensures a continuing, adequate supply of permanent and seasonal labor" now and after the United Kingdom leaves the European Union in March 2019 (BBC News, 2018). Access to Eastern European workers in Europe's high-income countries, like US farmers' access to Mexican workers, is critical in the short run. However, in the long run, all of these countries face the prospect that fewer children will grow up to be hired farmworkers in labor-source countries as those countries develop.

The "Labor Conservation" Response

In the world of resources, the alternative to exploration and development is conservation. The same is true for farm labor. Energy conservation entails shifting to less energy-demanding production and consumption activities, and developing alternative energy sources and efficiency-enhancing technical change. Farm labor conservation involves shifting to less labor-intensive crops and raising worker productivity via mechanization and improved management. Consumers' demand for fresh fruits and vegetables will continue to increase as incomes rise around the world, limiting farmers' economic incentives to switch to less labor-intensive crops. This leaves one real option: farmers will have to produce more food with fewer workers, and new technologies will have to become available to make this possible. If we could step into then out of Dr. Who's Tardis, into a not-so-distant future, would we find robots in the fields instead of low-skilled farmworkers?[7] That is the question we ask in Chapter 9.

---

7. The Tardis is a time-traveling telephone booth in the long-running British series "Dr. Who." To learn more, see The Tardis, http://www.thedoctorwhosite.co.uk/tardis/.

REFERENCES

BBC News, 2018. Food and Farming Sector Makes Post-Brexit Demands. BBC News. May 28, 2018. http://www.bbc.com/news/uk-politics-44274022. [Accessed 2 June 2018].

Charlton, D., Taylor, J.E., 2016. A declining farm workforce: analysis of panel data from rural Mexico. Am. J. Agric. Econ. 98 (4), 1158–1180.

El Colegio de la Frontera Norte, et al., 2012. Encuesta sobre Migración en la Frontera Sur de México (EMIF Sur). In: Indicadores Trimestrales de Coyuntura de la EMIF Sur, 2010–2012. www.colef.net/emif/resultados/indicadores/indicadores/Indicadores%20EMIFSUR%20I-2012.pdf.

Gabbard, S., Hernandez, T., MacDonald, S., Carroll, D., 2015. Changes in farmworker charateristics. In: Presentation at *Farm Labor and the ALRA at 40,* Davis, CA. April 17, 2015.

Handa, S., 2002. Raising primary school enrolment in developing countries: the relative importance of supply and demand. J. Dev. Econ. 69 (1), 103–128.

INM, 2012. Mexico, "El INM Favorece el Ingreso Documentado de Más de 29 Mil Trabajadores Fronterizos" (Boletín No. 002/12, January 3, 2012). www.inm.gob.mx/index.php/page/Boletin_0212.

Instituto Nacional de Migración (INM), 2011. Mexico, "México Documentó a Más de 26 Mil Trabajadores Fronterizos" (press release, January 3, 2011). www.inm.gob.mx/index.php/page/Noticia2_030111.

Marosi, 2015. Baja farmworkers win raises, benefits in landmark Deal. In: Los Angeles Times. June 5.

Martin, Philip. 2017. "Th H-2A Farm Geustworker Program is Expanding Rapidly: Here are the Numbers You Need to Know." *Economic Policy Institute*. Epi.org. April 13, 2017. Accessed 4 June 2018.

Martin, P.L., Fix, M., Taylor, J.E., 2006. The New Rural Poverty—Agriculture and Immigration in California. The Urban Institute Press, Washington, DC.

Massey, D.S., Taylor, J.E.e., 2004. International Migration: Prospects and Policies in a Global Market. OUP Oxford.

Ruttan, V.W., Hayami, Y., 1984. Toward a theory of induced institutional innovation. J. Dev. Stud. 20 (4), 203–223.

Smith, A., Taylor, J.E., 2017. Essentials of Applied Econometrics. University of California Press, Oakland, CA.

Taylor, J.E., 2010. Agricultural labor and migration policy. Annu. Rev. Resour. Econ. 2 (1), 369–393.

Taylor, J.E., López-Feldman, A., 2007. Does Migration Make Rural Households more Productive? Evidence from Mexico. ESA Working Paper No. 07-10, Agricultural and Development Economics Division (ESA), FAO, Rome.

Taylor, J.E., Charlton, D., Yúnez-Naude, A., 2012. The end of farm labor abundance. Appl. Econ. Perspect. Policy 34 (4), 587–598.

Terrazas, A., Papademetriou, D.G., Rosenblum, M.R., 2011. Evolving Demographic and Human-Capital Trends in Mexico and Central America and their Implications for Regional Migration. Migration Policy Institute, Washington, DC. www.migrationpolicy.org/pubs/RMSG-human-capital.pdf.

UN-ECLAC, 2011. Subregión Norte de América Latina y El Caribe: Información del Sector Agropecuario, 2000–2010. Chart 7. www.eclac.org/publicaciones/xml/6/44886/2011-060-Inf.sect.agrop.2000-2010-L.1040-1.pdf.

United States Department of Agriculture, Economic Research Service, 2016. Farm Labor. Last updated July 12. http://www.ers.usda.gov/topics/farm-economy/farm-labor/background.aspx.

United States Department of Labor, 2016. Findings from the National Agricultural Workers Survey (NAWS) 2013–2014. Washington, DC. https://www.doleta.gov/naws/pages/research/docs/NAWS_Research_Report_12.pdf.

FURTHER READING

Lewis, W.A., 1954. Economic development with unlimited supplies of labour. Manch. Sch. 22 (2), 139–191.

Martin, P., Taylor, J.E., 2003. Farm employment, immigration, and poverty: a structural analysis. Agric. Resour. Econ. Rev. 28 (2), 349–363.

Passel, J., Cohn, D., Gonzalez-Barrera, A., 2012. Net Migration from Mexico Falls to Zero- and Perhaps Less. PEW Hispanic Center, Washington, DC. Available at http://www.pewhispanic.org/2012/04/23/net-migration-from-mexicofalls-to-zero-and-perhaps-less/.

Chapter 9

Robots in the Fields

Farming robots are about to take over our farms.

Investor's Business Daily (10 August 2018)

Farmers confronting labor scarcity have three options: to seek workers in new places, to shift out of labor-intensive crops, or to produce more food with fewer workers by adopting labor-saving technologies. The agricultural transformation in migrant-sending countries and politics in destination countries limits the first option. Rising consumer demand for fresh, locally produced fruits and vegetables limits the second. In the future, staying competitive will require developing and using new technologies to make a smaller and more expensive farm workforce more productive. This chapter explains the economics behind the development and adoption of new agricultural technologies. A tekked-up agriculture will require tekked-up workers capable of using new technologies. Agricultural producers, workers, and rural communities will need to prepare for a world of farm labor scarcity, with robots in the fields.

In the mid-1950s, two scientists at the University of California, Davis, Coby Lorenzen, an engineer, and Jack Hanna, a plant breeder, joined forces to invent a machine that could harvest tomatoes. Few people thought they could do it, and even more wondered "why bother?" Farm labor was inexpensive. There were hundreds of thousands of Bracero workers, and even more undocumented immigrants willing to pick tomatoes alongside them. But the Bracero program was controversial and there were political pressures to end it. The Catholic Church opposed it on the grounds that it separated Mexican families and exposed migrant workers to vice. Powerful labor unions opposed it, arguing that it adversely affected the wages of domestic farm workers (Craig, 1971).[1] If the US government did end the Bracero program, it was clear that there would not be nearly enough domestic workers to take the Braceros' place. Besides, imagine the challenge of creating a mechanical beast that could move through the fields, picking fragile tomatoes without damaging them (at least not much), while separating each one from the vine?

---

1. Recent economic research suggests that excluding Bracero workers did not raise agricultural wages or substantially raise employment for domestic workers; see Clemens et al. (2018).

The Farm Labor Problem. https://doi.org/10.1016/B978-0-12-816409-9.00009-4

Early results were not encouraging. The tomatoes split, spurting juice into the fields as the machine pierced their tender skin. The machines hit dirt clods and broke down. Can a machine ever substitute for a gentle human hand? Jack Hanna and Coby Lorenzen thought so. If you cannot make the beast's fingers gentle enough, try breeding the tomatoes to be more thick skinned.

The researchers' breakthrough came just in time for the end of the Bracero program. It combined engineering and plant science in a new way. The vf-145 "UC tomato," tough and easy to de-stem, was invented to be picked by the UC tomato harvester. Farmers, worried about impending labor shortages, embraced the new machine, even though it was rickety and untried on the massive scale required to replace a vanishing Bracero workforce.

The new technology diffused across California's processing tomato farms almost overnight. Financiers and UC Cooperative Extension (CE) agents helped California's processing tomato farmers adopt the new technology. Within 5 years, virtually 100% of the processing tomato industry in the United States was using Hanna and Lorenzen's mechanical harvester, and most farmers were planting vf-145. Over the next 35 years, harvest labor requirements per ton of processing tomatoes dropped by 92%, while the US processing tomato harvest more than doubled, from 4.1 to 9.4 million tons (Fig. 9.1) (Thompson and Blank, 2000).[2]

In the wake of the Bracero Program and farm worker union activity, there were increases in labor productivity in other crops, as well, and farm wages began to rise faster than nonfarm wages. Some economists predicted that mechanical harvesting would replace hand harvesting in the United States "within a decade" (Cargill and Rossmiller, 1970; Martin, 2009).

Not all innovation was technological. Innovations in labor management enabled groups of farmers to maintain production with fewer workers, paying them higher earnings and providing more generous benefits. The showcase study, examined by Mamer and Rosedale (1980), is that of the Coastal Growers Association (CGA). The CGA recruited workers and synchronized their activities across member farms. It managed to reduce its number of pickers from 8,517 in 1965 to 1,292 in 1978, while paying an average hourly wage ($5.63) that was more than twice the minimum wage at the time ($2.65) and offering farmworkers benefits that included health insurance, paid vacations, and subsidized housing.

THE UC TOMATO HARVESTER

The UC tomato harvester was a raging success from a technological and production economics point of view, but all was not well in California's tomato

---

2. Tomato harvest data are from the US Department of Agriculture's Economics, Statistics and Market Information System, http://usda.mannlib.cornell.edu/MannUsda/viewDocumentInfo.do?documentID=1210, Accessed 1 June 2018.

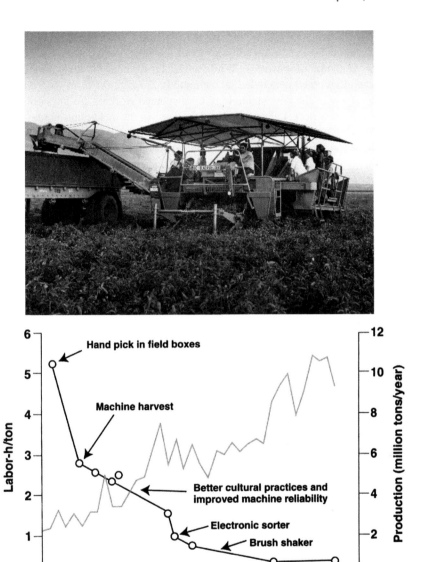

FIG. 9.1 The UC tomato harvester was introduced in 1962. Four years later, machines harvested 95% of the processing tomato crop. Improvements lowered labor use over time. *(Photo courtesy of UC Davis Library Special collections, Blackwelder Manufacturing Records D-326. Figure reproduced from Thompson, J. F., & Blank, S. C. (2000). "Harvest mechanization helps agriculture remain competitive." California Agriculture, 54(3), 51–56. https://doi.org/ 10.3733/ca.v054n03p51. Copyright © 2000 The Regents of the University of California. Used by permission.)*

fields. Small farmers could not afford the machines, nor could they compete with the large farmers who adopted them. Within 5 years, 88% of tomato farmers went out of business. While large farms consolidated around the new technology, they demanded an estimated 32,000 fewer farm workers.[3]

The effects of the new technology on displaced workers are impossible to measure in retrospect. Labor demand decreased in the fields but increased in processing plants that turned a growing volume of tomatoes into catsup, sauces, and other food products for an exploding fast-food industry. How difficult or easy was it for displaced tomato workers to find new jobs? How did tomato workers' new job opportunities compare with working in the tomato fields in terms of earnings? How did workers view the nonpecuniary characteristics of processing jobs compared with tomato harvesting? We do not have answers to these questions, because they were not a focus of economic research at the time. Nor was the impact on rural communities.

The California Agrarian Action Project, a group of small farmer and community activists, together with 19 farm workers, sued the University of California in 1979. They alleged that "the University's agricultural research program emphasized farm mechanization, which favored the interests of large agricultural business, to the detriment of the small farmer and consumer" and that "publicly funded mechanization research displaces farm workers, eliminates small farmers, hurts consumers, impairs the quality of rural life, and impedes collective bargaining."[4]

An analysis by Martin and Olmstead (1985) cast doubt on the validity of these charges, concluding that "applied research by universities is often authorized by legislation stipulating multiple goals, leaving researchers and universities vulnerable to lawsuits alleging that only some of the legislative goals are being pursued." Schmitz and Seckler (1970, p. 569) found that gross social gains from research and development expenditures on the tomato harvester were "in the vicinity of 1,000%."

The case kicked around in the courts for a decade. It eventually settled, but at a high cost to the University of California. Never before had there been such a challenge to mechanization research at a major US land-grant university. At a press conference in Fresno, California in December 1979, the US Secretary of Agriculture Robert Bergland stated "I will not put federal money into any project that reduces the need for farm labor" (Sarig et al., 2000).

---

3. University of California, Davis, Department of Plant Sciences, https://news.plantsciences.ucdavis.edu/2015/07/24/how-the-mechanical-tomato-harvester-prompted-the-food-movement/, Accessed 1 June 2018.
4. Court of Appeal, First District, Division 5, California, CALIFORNIA AGRARIAN ACTION PROJECT, INC. et al., v. REGENTS OF the UNIVERSITY OF CALIFORNIA et al., Decided May 25, 1989, https://caselaw.findlaw.com/ca-court-of-appeal/1760043.html, Accessed 1 June 2018.

As we saw in Chapter 6, most farmers had access to an abundant supply of foreign agricultural workers at low wages after the Bracero Program ended. The new wave of illegal immigration that helped derail unionization (Chapter 7) also discouraged mechanization. A survey of California farm workers by Mines and Martin (1986) found that unauthorized immigrants constituted at least one-fourth of the California farm work force in 1983. Using the same data, Taylor (1992) found evidence that the California farm labor market was segmented along legal status lines. In all, 10% of workers in relatively high-skilled "primary" jobs (e.g., machine operators) were unauthorized immigrants, compared with 32% in "secondary" jobs (harvesting and other manual labor). An econometric analysis found that unauthorized immigrants were more likely to be selected into low-wage, low-skill farm jobs than otherwise similar legal workers. Controlling for this selection process, the earnings of unauthorized workers were significantly lower than the earnings of similar legal workers in high-skill farm jobs. We saw in Chapter 8 that the SAW legalization program provided an additional bump up in the supply of low-skilled farmworkers after 1987.

In an environment of abundant farm labor, there was little incentive for farmers to adopt new labor-saving technologies or for universities and private companies to invest in the research and development needed to invent these technologies. The United States lagged behind other high-income countries in developing and manufacturing labor-saving crop technologies. When California vintners began using mechanical wine grape harvesters, they had to import the machines from France, at a price of $100,000–$300,000 each (Goldfarb, 2008).

Today, in an era of farm labor scarcity, farmers, universities, and technology developers are making up for lost time. What Hanna and Lorenzen did was unique in combining mechanical engineering with plant science. New labor-saving solutions combine these fields with computer science, including artificial intelligence and machine learning, and in some cases even chemistry. Many of the companies that are developing these technologies are in or near the nation's high-tech centers, including California's Silicon Valley. As these new labor-saving technologies become available, farmers will have to decide whether or not to adopt them. They will have to base their decision on a cost-benefit analysis, weighing the expected benefits, in terms of lower labor costs, with the costs, including possible adverse effects on product quality, lower harvest shares, and the up-front costs of adoption, which for some technologies are considerable.

MODELING THE ECONOMICS OF LABOR-SAVING TECHNOLOGY ADOPTION

We can adapt a mover-stayer model analogous to the one we learned in Chapter 3 to model farmers' decisions to stay with their current technology or move to a new one. The existing and new technologies are associated with different per acre production costs. Based on input prices, a farmer can optimize for each technology by finding the combination of inputs that minimizes the cost of producing a given level of desired output.

As wages rise, regardless of the technology a farmer uses, there are incentives to find substitutes for labor. For example, a farmer might try to hire fewer workers but provide each worker with more capital like power-assisted shears, moving platforms instead of ladders in orchards, or raised-bed ("tabletop") cultivation of strawberries, so that it is easier for workers to reach the fruit. These are examples of keeping technology relatively the same but shifting the labor-capital mix.

Alternatively, farmers can choose to adopt an entirely new technology that is "on the shelf." In economics, production functions represent the relationship between inputs and output. Technological change entails replacing one production function with an entirely different one. The tomato harvester is a textbook example of technology change. It replaced a harvesting process that was very labor-intensive with one that used hardly any labor at all. The same tonnage of tomatoes could be harvested with a fraction of the labor force.

A tomato harvester is a form of capital, but not one that was in the production function before. Switching to the new vf-145 tomato was part of this technology change. The workers changed, as well, from hand pickers to workers skilled at operating and maintaining the new technology.

When Henry Ford introduced the assembly line, it suddenly was possible to produce many more cars with less labor. Factories changed, becoming more horizontal to accommodate the assembly line. A similar transformation can occur in agriculture, as farmers invest to reconfigure orchards and fields around new technologies. New agricultural technologies may require planting new varieties of trees, vines, and field crops; erecting new trellis systems; and spacing and repositioning plants and trees.

The adoption of labor-saving technologies can entail considerable costs. For a farmer to adopt, the economic benefits have to be greater than the costs of adoption. New technologies that are not economically feasible when wages are low suddenly can become profitable once wages rise. Uncertainty about finding sufficient workers to harvest crops can add to the economic benefits from new technologies. Part of the benefit a harvesting machine can offer is insurance against a labor shortage, and the farmer might view part of the adoption cost as a sort of "labor-shortage insurance premium."

Imagine a farm that starts out with a labor-intensive technology described by the production function $Q^0 = F^0(L, K)$, where the "0" superscripts denote "technology 0," Q^0 is output, and L and K are labor and capital inputs, respectively. We assume that this production function has the usual properties, including positive marginal products of labor and capital and diminishing marginal returns to each factor.

If we were to graph this production function, it would be a surface in a three-dimensional (3D) space with axes for L, K and a vertical axis for output Q^0. If we

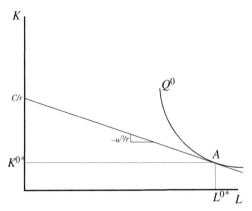

FIG. 9.2 A cost-minimizing farm using a labor-intensive technology optimizes at point A, demanding L^{0*} units of labor and K^{0*} units of capital.

choose some level of output and take a perfectly horizontal slice of this production function in the L, K direction we can graph it in the two-dimensional (2D) L, K plane. This "slice" is an isoquant curve like the one labeled Q^0 in Fig. 9.2. The isoquant curve shows all combinations of L and K that can generate the selected quantity of output Q^0 efficiently with technology 0. The isoquant in this figure reflects a technology that uses large amounts of labor relative to capital to produce the crop, not unlike most fruit, vegetable, and horticultural (FVH) production in the world today. You can tell that it is a labor-intensive technology because it lies so far toward the southeast corner in this figure.

What combination of labor and capital will the farm use to produce this given level of output with technology 0? To answer that question, we need to bring in the prices of labor (the wage, w) and capital (the rental rate, r). The farm's total cost of production is

$$C = wL + rK$$

An isocost line is the set of combinations of L and K that have the same cost, C. To derive isocost lines, just solve the cost equation for K as a function of C and L, since K is on the vertical axis of the graph in Fig. 9.2:

$$K = C/r - (w/r)L$$

The intercept C/r positions the isocost line on the vertical (K) axis. The slope of the isocost line is the negative ratio of the wage, w, to the rental rate on capital, r. There is an infinite number of possible isocost lines, one for each level of total cost, C. A decrease in total cost, other things being the same, shifts the intercept of the isocost line downward, keeping the slope of the line unchanged. That is, for a given rental rate r and wage w, it shifts the isocost line in toward the origin.

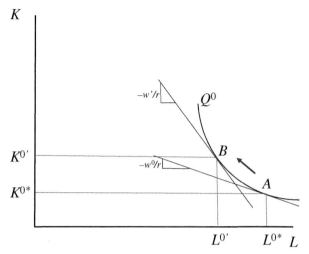

FIG. 9.3 With the labor-intensive technology, a wage increase induces the farm to shift its input mix from point A to point B, demanding more capital and less labor. The cost of producing output Q^0 increases due to the higher wage.

At initial wage w^0, the lowest cost way to produce output Q^0 is given by point A in Fig. 9.2. This is the point of tangency between the isoquant curve and the lowest isocost line that is feasible to produce Q^0. The lowest feasible isocost line just touches the isoquant curve. At this point, the slope of the isoquant curve equals the slope of the isocost line. The absolute value of the slope of the isoquant curve is the ratio of the marginal product of labor (MPL) to the marginal product of capital (MPK), MPL/MPK. The slope of the isocost line (in absolute value) is the ratio of w^0 to r. In words, the farm optimizes by choosing L and K so that the last dollar spent on L adds the same amount to output as the last dollar spent on K. In Fig. 9.2, the optimizing farmer with a labor-intensive technology hires L^{0*} units of labor and K^{0*} units of capital to product a level of output equal to Q^0.

What if labor becomes more expensive in an era of farm labor scarcity? The wage increases relative to the price of capital. This makes the isocost line steeper, as shown in Fig. 9.3. If the farm uses the same technology as before, at higher wage w' it will re-optimize by moving from point A to point B along isoquant Q^0, but its cost of producing Q^0 will increase. (By extending the isocost lines until they hit the vertical axis, it is clear that C is higher than it was before.) At point B, the farm demands less labor and more capital. Its cost of producing the same level of output as before rises because one of its inputs is more costly now. The farm does not have very much flexibility to adjust its input mix as long as it is stuck on the same isoquant.

Now suppose that a university and/or private company invents a new labor-saving technology, like the UC tomato harvester. With this technology, which

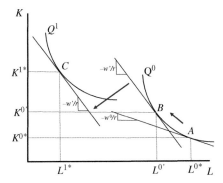

FIG. 9.4 Given the option of adopting a labor-saving technology, the farm depicted in this diagram will do it—provided that the cost savings exceed the (annualized) sunk cost of adoption. We can calculate the minimal cost of production with each technology from the intercepts of the isocost lines tangent to the isoquants. The intercept of the isocost line tangent to Q^1 is C^1/r, and the intercept of the isocost line tangent to Q^0 is C^0/r. (We would need a tall graph to show C^0/r!)

we will call technology 1, the farm can produce the same level of output as before with substantially less labor, which is expensive, and more capital, which is cheap relative to labor. The isoquant corresponding to this new technology is the curve labeled Q^1 in Fig. 9.4.

To understand this figure, it is important to keep in mind that both isoquant curves, Q^1 and Q^0, correspond to *the same level of output*. The difference between them is that Q^1 embodies a labor-saving technology (like the tomato harvesting machine, which requires costly capital investments and little labor) whereas Q^0 uses a labor-intensive technology (hand picking). Both produce the same tonnage of tomatoes.

Given the option of switching to the new technology, will the farmer do it? The answer depends mostly on two things: first, the impact on the cost of producing the same quantity of the crop as before, and second, the cost of switching from the old to the new technology.

We can see the effect on the total cost by comparing the two isocost lines with slope equal to the ratio of the absolute value of the new wage and rental rate, w'/r. The labor-saving technology shifts the isocost line to the left. Recall that the intercept of an isocost line is C/r. When the isocost line shifts leftward, the intercept falls. Assuming the rental rate on capital does not change, total cost decreases: $C^1 - C^0 < 0$. If this decrease in cost of production exceeds the cost of adopting the new technology, appropriately amortized as an annual cost of adoption, our model predicts that, given the option, the farmer will switch to the new technology when farmworker wages increase. (A more formal treatment of the technology adoption cost-benefit calculus appears in this chapter's Appendix.)

Agricultural economists Vernon Ruttan and Yujiro Hayami argued that, when relative factor prices change, the potential gains from new technologies induce research and development, a process they called *induced innovation* (Ruttan and Hayami, 1984). When the potential gains from technological change are large, private companies have a profit incentive to invest in research and development of new technologies. Farmers have incentives to pressure universities and other public research institutions to do the same.

The rest of this chapter describes a few of the most recent responses to a declining farm labor supply and advances in developing labor-saving technologies for agriculture in an era of farm labor scarcity.

RESPONSES TO A SHRINKING FARM LABOR SUPPLY[5]

The end of farm labor abundance has far-reaching implications for US and Mexican farmers, who can no longer count on having a steady supply of low-wage workers. Immigration will not be the answer to meeting US farm labor demand in the future. Rising wages already have stimulated a search for alternatives on US farms, including increased use of the H-2A contract worker program, changes in crop mix, and the development and adoption of new technologies.

Agricultural producers are adopting more efficient labor management practices, such as hiring foreign guest workers through labor and growers' associations that streamline the hiring process and educate employers on laws, regulations, and best practices. One such association is WAFLA (f.k.a. Washington Farm Labor Association; https://www.wafla.org/), which offers services to help farm employers file for H-2A visas, recruit workers, and comply with the lodging, transportation, and other requirements of the H-2A program. It initially served large farms, but increasingly it recruits workers for small farms, as well.

Some producer organizations, including the North Carolina Growers Association and Virginia Agricultural Growers Association, serve as joint employers with growers instead of merely hiring agents, so workers can move freely to member farms where they are most needed during their visa term. Efficient labor management practices, along with maintaining strong connections with a shrinking foreign workforce, become more vital as the agricultural transition unfolds in Mexico and other labor-source countries.

Guest worker programs are only a short-term option, at best: they have to recruit from an ever-smaller pool of available farm workers. There is little hope of recruiting enough farm workers from other nations to take the place of Mexican workers on US farms. The challenge is complicated by the fact that Canadian farms also recruit farm workers from Mexico, which in turn is recruiting farm workers from Guatemala. The Central American farm workforce is far too small to meet the labor demands of the United States, Canada, Mexico, and Central America. The total rural population of Central America was 6.9 million smaller than the rural population of Mexico in 2016 (The World Bank, 2017). H-2A workers are being recruited from a larger diversity of places than

---

5. The following two sections draw heavily from Charlton et al. (2018).

previously. However, it is not reasonable to expect the US government to expand this guest worker program on a large enough scale to come close to filling the void left by Mexican workers leaving agriculture.

US farmers could respond to farm labor shortages by growing fewer fresh fruits and vegetables that require large amounts of labor. To a certain extent, we observe this occurring. Acreage of handpicked fruits like apples, peaches, and citrus has decreased in the United States since 2002 while acreage of mechanically harvested almonds has increased. Nevertheless, niche markets for locally grown produce have expanded in many communities of the United States. As long as consumer demand for domestically grown fresh fruits and vegetables persists, we will continue to see these commodities grown in the United States.

A more viable option is to invest in the development and adoption of labor-saving technologies to meet the demands of consumers in the midst of a declining farm labor supply. The big question for farmers, researchers, and policymakers is: Will new labor-saving technologies keep pace with the shrinking farm labor supply, and what are the implications for farm workers and the communities in which they live?

A SURGE IN AGRICULTURAL INNOVATION[6]

Picking a ripe peach or strawberry is trivial for humans. Our brains work seamlessly with our hands, and human fingers are custom made to delicately pick even the most fragile fruits and vegetables. Developing agricultural technologies to replace human hands is challenging, because the technologies must be discerning, quick, and dexterous (Vougioukas, 2016).

The human eye can discern unripe from ripe fruit and leave the unripe fruit on the vine or tree to mature. Photographic sensors are developing much the same skill. This can enable a robot to mimic the discernment that a picker can offer. Accuracy of discernment must also be combined with speed. Human harvesters can train themselves to discern and pick at high speeds, but how quickly can a machine detect ripe fruit and pick it? Even with discernment and speed, machines face the challenge that most fruits and vegetables are delicate and require a high level of dexterity to pick the fruit without bruising or crushing it.

It is one thing for engineers to create a robot with one or two of these characteristics needed to replace human harvesters, but it is more difficult to implement all three.

Fruits that will be processed into juice, sauces or preserves have been mechanically harvested for many years; the tomato harvester is an example. Another is raisin grapes. California's San Joaquin Valley around the city of Fresno produces up to one-third of the global supply of raisins. To do this, it traditionally employed 40,000–50,000 workers for a few weeks each year to cut down bunches of grapes with a sharp knife, collect them in a tub, and lay

---

6. This section draws from Charlton et al. (2018).

them on paper trays to dry into raisins. Farmers do not hire workers until the sugar level in the grapes reaches 20%–24%, but then harvesters are required immediately to maximize the drying time before the September rains come (Martin et al., 2006, p. 36). Farmers complain of labor shortages even in the best years. Hand-harvesting grapes for raisins is extremely labor intensive, requiring 19 person-hours per ton of dry raisins (Christensen, 2000).

The University of California's Cooperative Extension (CE), whose cost and returns studies provided the data we used to model farm labor demand in Chapter 2, describes the traditional method in its study of tray-dried raisins (Peacock et al. 2006):

> *Harvest consists of hand picking the grapes into pans. Paper trays are placed by the picker on the upper one-half of the terrace and the grapes are spread evenly on the paper trays...Raisins are rolled at 16%–18% moisture, allowed to equilibrate and then boxed when moisture is 14% or less...The crop is dumped into bins that hold 1,000 to 1,200 pounds of raisins, a process referred to as boxing...Papers are burned at the end of the row when weather conditions permit.*

Labor demands in the raisin harvest can be reduced to only three person-hours per ton by using dry-on-vine (DOV) technology. The CE describes the DOV harvesting method as follows (Fidelibus et al., 2016):

> *Canes bearing fruit are severed in August to allow the fruit to dry on the vine. The custom harvesting operation brings along with the harvester two bin trailers, two tractors, one flatbed truck and a forklift. The over-the-row harvester picks all the fruit in one pass per vine row. The crop is harvested into one-half ton bins carried on the harvester which includes the harvester driver and an assistant.*

Despite the labor savings from DOV, in 2000, almost all of the 270,000 acres of raisins grown in California were still harvested by hand, and in 2016, only 9% of acreage was harvested using DOV (Fidelibus, 2014; NASS, 2016).

Changes in production practices and high up-front costs are the main reason for the slow adoption of labor-saving technologies in raisin production. DOV requires planting an entirely different variety of grape, installing a new trellis system, and purchasing a mechanical harvester. This means that adoption is only optimal when farm wages are sufficiently high and producers can use the newly adopted technology over a long time horizon to justify the investment. In 2002, a 56% drop in raisin prices and a tightening labor market led raisin producers to adopt labor-saving technologies more quickly, and DOV acreage increased (Fidelibus, 2014).

Raisin grapes have the advantage that, with DOV, they can be set to dry in the sun without any person or machine touching the ripe fruit, and once dried, the raisins do not require delicate handling. Fresh-to-market fruits require an entirely different level of dexterity to harvest. Consumers will not notice some bruising on tart cherries processed in sugar and frozen or canned for pie filling.

Thus, farms can harvest tart cherries by shaking the cherries from the trees and catching them in a frame. However, bruises and broken skin would be unacceptable on sweet cherries that are eaten fresh. The mechanics of harvesting tart cherries implements speed. Little discernment is needed as long as cherries ripen uniformly on the tree (which agronomic research can assist), and dexterity is nonexistent in shake-and-catch systems.

Strawberries are a different story. Robotic strawberry harvesters are under development, but the technologies they must embody are complex. These harvesters would use graphic processing to determine whether the strawberry is red and ripe for picking, pick the strawberry with a robotic arm that grabs and cuts the stem without touching the berry, and pack strawberries into baskets and trays (Agrobot, 2018). Strawberry-picking robots under development are not as fast as a human hand, but they can run day and night to make up for lost time. At the present stage of development, robots still have difficulty detecting ripe berries that are hidden under a mass of leaves. Robots being developed by the company Harvest CROO robotics were able to detect and harvest about 50% of ripe berries in 2018. That is a poor record compared with a human crew that picks 60%–90%. Nevertheless, CROO believes they will have a feasible machine on the market in 2 years (Charles, 2018).

Mechanical technologies are reaching areas of agriculture that they never reached before, including automated lettuce thinners, integrated weed management systems that recognize and selectively spray weeds, and other uses of precision technology. Lettuce, in order to form into large heads, must be thinned to 10 inches apart. Conventional thinning involves workers with hoes walking through the fields, deciding which lettuce starts to keep and taking out the rest. Automated lettuce thinners combine mechanical engineering, machine learning, and chemistry to identify which starts to keep and knock out the rest with a jet spray of concentrated fertilizer. This reduces labor demands from 7.31 person-hours per acre to only 2.03 (Mosqueda et al., 2017). Robotic weeding technologies use similar machine-learning methods to distinguish weeds from productive plants then target the weeds with miniscule jets of herbicide. Precision technologies not only reduce labor hours; they also can reduce the use of pesticides, fertilizers, and other inputs. Precision technologies that reduce chemical use can be beneficial to farm workers' health and the environment.

Unlike FVH production, dairies require a steady labor supply year round. US dairies employ an average of four full-time workers and 1.6 part-time workers for an average herd size of 297 cows (Rosson, 2012).[7] In all, 47% of hired workers on dairies are foreign, and labor shortages are a common concern.

Robotic milking systems (RMSs) can reduce labor demand, but reports of actual labor savings and profit potentials vary widely. This technology became commercially available in the early 1990s. It uses a milking machine, laser

---

7. This study dropped all dairies with fewer than 50 milk cows. The mean number of milk cows for the remaining dairies was 297.

sensors, robotic arms, and a gate system that controls when cows are milked. Some farmers allow cows to choose when to be milked, while other farmers herd the milk cows through the milking parlor at preset times of day.

RMS impacts on milk production range from losses of 5%–10% to gains of 5%–10% or 15%–20% compared with conventional milking (Salfer et al., 2017; Floridi et al., 2013). There were 35,000 RMS units in use around the world in 2017 (Salfer et al., 2017). Adoption is more common in Europe than in North America. It may become more widespread as farmers find more efficient ways to utilize the labor-saving potential of RMS and better interpret and use data collected from milking machines to improve quantity and quality in milk production.

Some innovations are further from becoming feasible for widespread adoption. For example, engineers are adopting apple-harvesting robots that locate and pick fruit using computer vision technology and vacuum grips on robotic arms. Prototypes have been tested in V-trellised trees so that the robot only picks from a 2D space. Results are encouraging, with the robot picking one apple per second compared with a human picking one apple per 1.5 seconds (Salisbury and Steere, 2017). This technology requires redesigning the orchard around the harvester, and the V-trellis reduces fruit per tree. Engineers are currently seeking to develop robotic harvesters that can harvest fruit within the canopy of the tree more efficiently and delicate fruits like peaches and cherries without bruising or damaging them.

CONCLUSION

There are many technological possibilities for responding to worker shortages in an era of diminishing farm labor supply. A selection of technologies, from established ones to early prototypes, can be found on the website for this book.[8]

Whether or not farmers adopt new technologies depends on the success of science and industry to create affordable and reliable technologies to replace humans, but also on the availability and cost of hiring workers to do manual labor. As we saw in Chapter 8, few people in high-income countries are willing to work in agriculture, and the workforce in less-developed countries is transitioning out of agriculture. This means that robots might well be the physical "workers" in the fields soon, and growers will be hiring engineers, mechanics, and technicians to develop, operate, and repair them.

New technologies make farm workers more productive, and this makes it possible for farmers to pay higher wages to a smaller workforce. Rising wages can benefit farm workers and the communities where they live, but only if workers have the skills that new technologies demand, and if lower-skilled workers can retool so they can be part of more mechanized production

---

8. farmlabor.ucdavis.edu

processes. Innovations that keep an aging farm workforce employed and productive are needed while researchers develop robots that can perform tasks that seem simple for humans but are challenging for machines. Some innovations make use of relatively simple technologies, like growing berries on platforms in the fields that save workers' backs or providing workers with power-assisted pruning shears. A diminishing farm labor supply puts pressure on the agricultural sector to adopt new technologies for difficult-to-mechanize tasks. The competitiveness of agriculture, as well as the welfare of farm workers and the communities in which they live, depend on how we as a society adapt to an era of farm labor scarcity.

APPENDIX

Cost-Benefit Analysis of Adopting a Labor Saving Technology: Raisin Grapes

In this chapter, we learned that if a labor-saving technology decreases the cost of production, and this decrease exceeds the cost of adopting the new technology, amortized as an annual cost of adoption, farmers will switch to the new technology. Other things being the same, farmers will be more likely to adopt a labor-saving technology when farmworker wages increase. This Appendix offers a more detailed and formal approach to analyze the costs and benefits of adopting labor-saving technologies.

We can analyze technology adoption using a mover-stayer model, because the farmer's decision is to either "stay" with the current technology or "move" to a new one. Recall from Fig. 9.3 that the cost-minimizing combination of labor and capital with the existing technology after the wage rises from w^0 to w' is $(L^{0'}, K^{0'})$ per year. This means that the total annual cost of producing the given level of output if the grower sticks with the old technology is

$$C^0 = w'L^{0'} + rK^{0'}$$

The optimal combination of labor and capital to produce the same output with the labor-saving technology (Fig. 9.4) is (L^{1*}, K^{1*}). So, the lowest total annual cost using the labor-saving technology is

$$C^1 = w^1L^{1*} + rK^{1*}$$

The annual cost savings from adopting the new technology are $C^0 - C^1$. In the example given in Fig. 9.4, the labor-saving technology results in a lower total cost, so this difference is positive. The cost savings are the benefit from adopting the new technology in this mover-stayer model.

To see whether or not it is optimal for the farmer to adopt, we need to compare the benefit in terms of annual cost savings with the annualized cost of adoption. The annualized cost of adoption is the total cost of adoption amortized over the assumed life of the investment, which we can call n months. Let I denote the up-front investment required to adopt the new technology, and let the Greek letter δ (delta) denote this up-front cost amortized over n months. We can think of δ as the annual payment of principal plus interest that the farmer would make on a loan equal to the full up-front cost of adopting the new technology. Three things are needed to calculate δ:

- The total investment required to switch to the new technology, I.
- The monthly interest rate, i, which is simply the annual interest rate/12.
- The number of months over which the investment will be amortized (the loan term), n.

With this information, we can calculate δ by using an online loan calculator or, alternatively, the formula the calculator uses, which is

$$\delta = 12^* I / D$$

D is the discount factor corresponding to the interest rate that would apply to a loan to finance the new technology and the loan term. It is calculated as

$$D = \frac{(1+i)^n - 1}{i(1+i)^n}$$

The analysis becomes slightly more complicated if the new technology has an impact on production as well as input demand. In the case of continuous-tray DOV raisin technology, output per acre is more than twice that of the conventional technology. We have to be careful to adjust for this difference in output per acre while carrying out the calculations in this Appendix. We do this by keeping output constant, like in Fig. 9.2, and calculating input and adoption costs per unit (ton) of output. We also have to divide δ by the per-acre yield, in tons.

Now we are ready to perform a cost-benefit analysis of the decision of whether or not to adopt a labor-saving technology. You can do your own cost-benefit analysis by using the "Technology Cost-Benefit Analysis" excel file on the website for this book. Letting $A_i = 1$ denote adoption by farmer i and $A_i = 0$ denote nonadoption, the decision rule is

$$A_i = \begin{cases} 1 & \text{if } w'L^{1*} + rK^{1*} + \delta < w^0 L^{0*} + rK^{0*} \\ 0 & \text{otherwise} \end{cases}$$

Economics of Technology Change in Labor-Intensive Raisin Grape Production

We can use the method above to calculate the cost and returns to converting a traditional tray-dried raisin vineyard to a DOV operation in the San Joaquin Valley of California. Sample costs come from Fidelibus et al. (2016) and Peacock et al. (2006), respectively.

Traditional tray-dried raisin production is highly labor intensive and traditionally uses different grape varieties and trellis systems than DOV production. Thompson Seedless vines are grown with 7×12 ft spacing and 519 vines per acre on our sample tray-dried raisin farm. The trellis system consists of two wires and a 24-in. cross-arm design. The trellis is considered as part of the vineyard, since it will be removed with the vines when the vines are no longer productive. Vines begin producing in the third year after planting and are expected to continue producing for another 22 years. Average yields are around 2 tons per acre in the fourth year and after.

Grapes are handpicked from mid-August to mid-September. The grower contracts a custom hand harvestor. One man can pick about one-third acre per 10-hour day, or 1 raisin ton per 15 hours, where one raisin ton is equivalent to 4.5 tons of fresh grapes. The grapes are handpicked into pans. Paper trays are placed next to the picker on the upper half of the terrace and grapes are spread evenly onto the paper trays with 18–20 pounds of fresh fruit on each tray. Raisin trays are rolled when the raisins reach 16%–18% moisture. The raisins are allowed to equilibrate and are boxed at 14% moisture or less. The grower then rents a tractor to pull a bin trailer and forklift for loading and unloading bins. Each bin holds 1,000–1,200 pounds of raisins. Labor for loading and unloading includes a tractor driver, someone to ride the bin trailer and remove paper trays, two people to pick up the rolled raisins and throw them into the bins, and a forklift operator to work the staging area, unloading and loading bins and transporting the loaded and empty trailers to and from the boxing crew.

DOV production reduces harvest operations because grapes dry on the vine rather than on trays placed between rows, but to switch from tray drying, producers must plant new vines and establish a new trellis system. The cost estimates in this example assume that a vineyard starts out already planted using the traditional technology and is then converted to DOV. To convert the land to DOV, the land is chiseled twice, leveled, and disked. A new trellis is installed, and vines are planted with 6×10 ft spacing, 726 vines per acre. Early maturing varieties must be planted to ensure that grapes will dry on the vine before fall rains come. Potential varieties include Fiesta or Selma Pete. The expected life of the vineyard is 30 years. Harvest begins in the third year with 2–3 tons per acre, and vines reach maturity in the fourth year, producing 5 tons per acre on average, more than twice that of continuous tray. To harvest, canes (stems) bearing fruit are severed in August. Once grapes have dried (on the vine), an over-the-row harvester picks all the fruit in one pass per vine row, harvesting the raisins directly into bins.

Sample costs for continuous tray and DOV appear in Table 9.A1. We converted all of the costs to 2016 dollars. Machine labor has a higher pay rate than hand labor, so we estimate machine and hand labor hours separately using information from the cost studies. We apply the same pay rates to both continuous tray and DOV for comparability. Since DOV has over twice the expected yield of continuous tray, we calculate costs per ton. This makes our analysis

TABLE 9.A1 Sample Costs and Yields for Traditional Tray-Dried and Mechanical Dry-on-the-Vine Raisin Production

Variable	Raisin Technology	
	Traditional Tray-dried	*Dry-on-the-Vine (DOV)*
Yield (tons per acre)	2.00	5.00
Hand labor wage (incl overhead)	$ 16.68	$ 16.68
Machine labor wage (incl overhead)	$ 22.24	$ 22.24
Hand labor hours per ton	39.82	15.20
Machine labor hours per ton	7.46	2.14
Labor cost per ton	$ 830.15	$ 301.20
Capital costs per ton	$ 75.60	$ 135.60
Other input costs per ton	$ 575.00	$ 114.20
Total annual operating cost per ton (C^0, C^1)	**$ 1480.75**	**$ 551.00**
Operating cost savings with DOV	**$**	**929.75**
Vineyard life (years)	22.00	27.00
Total establishment costs per acre		$ 13,994.00
Annual interest rate		0.0375
D		203.56
Annual establishment costs per ton (δ)		**$ 164.99**

compatible with the isoquant curve analysis in Fig. 9.4. We exclude the cost of land and property taxes, since we are considering the transition to DOV production on the same property.

The table shows that all annual operating costs per ton (labor, capital, and other inputs such as pesticides and herbicides) are lower with DOV than with the traditional technology. Labor costs are considerably lower for DOV: $301.20 compared with $830.15. Capital costs are higher: $135.60 vs. $75.60. Higher capital costs but lower labor costs are consistent with the two technologies pictured in Fig. 9.4. Other input costs are also lower with DOV than with the traditional tray-dried technology. A major source of savings is that DOV saves the cost of paper trays. The total annual operating cost per ton is $1,480.75 for the traditional technology and $551 for DOV. These numbers correspond to C^0 and C^1, respectively, in Fig. 9.4 and the cost equations at the start of this Appendix; however, they include other input costs as well as the costs of labor and capital. The difference in total annual operating cost per ton is $929.75, in favor of DOV.

In light of the savings in annual operating costs, should farmers adopt DOV? The answer to this question depends on the annualized up-front cost of switching from traditional to DOV raisin grape production.

The total sunk cost of establishing a DOV vineyard is $13,994 per acre.

To find the annual establishment costs per ton, we amortize the establishment cost per acre over 27 years of production life for the vineyard and divide by the yield. The annual interest rate we use is 3.75%, although of course rates vary over time and from farmer to farmer. The higher the interest rate, the less profitable it will be to switch to DOV. At an annual rate of 3.75%, the monthly interest rate is $i = \frac{.0375}{12} = 0.003125$. It gets compounded over a total of $n = 27 * 12 = 324$ months over the life of the vineyard. The annual payment with monthly compounded interest follows the formula:

$$12 * 13,994 / (D * \text{yield})$$

Using the formula given above, for an interest rate of 3.75% annually (0.003125 monthly) and 324 months, D is calculated as

$$D = \frac{(1 + .003125)^{324} - 1}{.003125(1 + .003125)^{324}} = 203.56$$

The yield with DOV is 5 tons/acre. Thus, the annual payment is

$$\delta = 12 * 13,994 / 203.56 * (1/5) = 164.99$$

The annualized sunk cost of converting to DOV is smaller than the annual operating-cost savings: $164.99 < 929.75$. In this example it would be profitable to replace the traditional tray-dried raisin vineyard with DOV even if vines planted under the traditional technology have many more productive years left.

REFERENCES

Agrobot, 2018. Agrobot Robotic Harvesters. Agrobot.com. (Accessed 4 June 2018).

Cargill, B.F., Rossmiller, G.E. (Eds.), 1970. Fruit and Vegetable Harvest Mechanization: Policy Implications. Rural Manpower Center, Michigan State University, East Lansing.

Charles, D., 2018. Robots are trying to pick strawberries. So far, They're not very good at it. In: All Things Considered. Npr.org. (Accessed 4 June 2018).

Charlton, D., Edward Taylor, J., Vougioukas, S., 2018. Can Wages Rise Quickly Enough to Keep Workers in the Fields? Montana State University Department of Agricultural Economics and Economics and the University of California, Davis, Departments of Agricultural and Resource Economics and Biological and Agricultural Engineering, respectively.

Christensen, L. (Ed.), 2000. Raisin Production Manual. vol. 3393. UCANR Publications.

Clemens, M.A., Lewis, E.G., Postel, H.M., 2018. Immigration restrictions as active labor market policy: evidence from the Mexican bracero exclusion. Am. Econ. Rev. 108 (6), 1468–1487.

Craig, R.B., 1971. The Bracero Program: Interest Groups and Foreign Policy. University of Texas Press, Austin.

Fidelibus, M., 2014. Grapevine cultivars, trellis systems, and mechanization of the California raisin industry. Hort Technol. 24 (3), 285–289.

Fidelibus, M., Ferry, L., Jordan, G., Zhuang, D., Sumner, D., Stewart, D., 2016. Sample Costs to Establish a Vineyard and Produce Dry-on-Vine Raisins—Open Gable Trellis System—Early Maturing Varieties—San Joaquin Valley. Univ. California Coop. Ext., Dept. Agr. Resource Econ. Davis.

Floridi, M., Bartolini, F., Peerlings, J., Polman, N., Viaggi, D., 2013. Modelling the adoption of automatic milking Systems in Noord-Holland. Bio-based and Appl. Econ. 2 (1), 73–90.

Goldfarb, A., 2008. Moving toward mechanical: High-end winemakers warm slowly to machine harvesting. In: Wines and Vines. (March). https://www.winesandvines.com/features/article/53452/Moving-Toward-Mechanical. (Accessed 1 June 2018).

Mamer, J., Rosedale, D., 1980. The Management of Seasonal Farm Workers Under CollectiveBargaining. University of California Cooperative Extension Leaflet 21147, March.

Martin, P.L., Olmstead, A.L., 1985. The agricultural mechanization controversy. Science 227 (4687), 601–606.

Martin, P.L., 2009. Importing Poverty?: Immigration and the Changing Face of Rural America. Yale University Press.

Martin, P.L., Fix, M., Taylor, J.E., 2006. The New Rural Poverty. The Urban Institute, Washington, DC.

Mines, R., Martin, P.L., 1986. A Profile of California Farmworkers, Giannini Information Service 86-2. University of California, Division of Agriculture and Natural Resources, Berkeley.

Mosqueda, E., Smith, R., Goorahoo, D., Shrestha, A., 2017. Automated lettuce thinners reduce labor requirements and increase speed of thinning. Calif. Agric. 1–6.

National Agricultural Statistics Service, USDA-CDFA, 2016. California Raisin Grape Mechanical Harvest Report. https://www.nass.usda.gov/Statistics_by_State/California/Publications/Crop_Releases/Acreage/2017/201705grptrelfinal.pdf. (Accessed 17 April 2018).

Peacock, W.L., Vasquez, S.J., Hashim, J.M., Fidelibus, M.W., Leavitt, G.M., Klonsky, K.M., De Moura, R.L., 2006. Sample Costs to Establish a Vineyard and Produce Grapes for Raisins: Tray Dried Raisins: San Joaquin Valley. University of California Cooperative Extension.

Rosson, C.P., 2012. Regional views on the role of immigrant labor on U.S. and southern dairies. J. Agric. Appl. Econ. 44 (3), 269–277.

Ruttan, V.W., Hayami, Y., 1984. Toward a theory of induced institutional innovation. J. Dev. Stud. 20 (4), 203–223.

Salfer, J., Endres, M., Lazarus, W., Minegishi, K., Berning, B., 2017. Dairy Robotic Milking Systems—What are the Economics? In: eXtension. October 3, 2017, eXtension.org. (Accessed 4 June 2018).

Salisbury, K., Steere, D., 2017. Phase 2 System Integration. Abundant Robotics, Inc. *Final project report,* Washington Tree Fruit Commission. http://jenny.tfrec.wsu.edu/wtfrc/core.php?rout=displtxt&start=1&cid=799. (Accessed 20 March 2018).

Sarig, Y., Thompson, J.F., Brown, G.K., 2000. Alternatives to Immigrant Labor? The Status of Fruit and Vegetable Harvest Mechanization in the United States. In: Center for Immigration Studies. http://www.cis.org/FarmMechanization-ImmigrationAlternative.

Schmitz, A., Seckler, D., 1970. Mechanized agriculture and social welfare: the case of the tomato harvester. Am. J. Agric. Econ. 52, 569–577.

Taylor, J.E., 1992. Earnings and mobility of legal and illegal immigrant workers in agriculture. 74 (4), 889–896.

The World Bank, 2017. World DataBank. http://databank.worldbank.org/data. Last checked 1 September 2017.

Thompson, J.F., Blank, S.C., 2000. Harvest mechanization helps agriculture remain competitive. Calif. Agric. 54 (3), 51–56.

Vougioukas, Stavros. 2016, Personal communication with the author. UC Davis, Department of Biological and Agricultural Engineering. April 20, 2016.

FURTHER READING

Richard, B., 1971. Craig, The Bracero Program: Interest Groups and Foreign Policy. University of Texas Press.

Vasquez, S., Fidelibus, M., Christensen, L., Peacock, W., Klonsky, K., DeMoura, R., 2007. Sample costs to produce raisins - Continuous tray-harvest equipment purchased used and refurbished, San Joaquin Valley. Univ. California Coop. Ext., Dept. Agr. Resource Econ. Davis.

Index

Note: Page numbers followed by *f* indicate figures and *t* indicate tables.